Differential Algebraic Topology
From Stratifolds to Exotic Spheres

Differential Algebraic Topology
From Stratifolds to Exotic Spheres

Matthias Kreck

Graduate Studies
in Mathematics
Volume 110

American Mathematical Society
Providence, Rhode Island

EDITORIAL COMMITTEE
David Cox (Chair)
Rafe Mazzeo
Martin Scharlemann

2000 *Mathematics Subject Classification.* Primary 55–01, 55R40, 57–01, 57R20, 57R55.

For additional information and updates on this book, visit
www.ams.org/bookpages/gsm-110

Library of Congress Cataloging-in-Publication Data
Kreck, Matthias, 1947–
  Differential algebraic topology : from stratifolds to exotic spheres / Matthias Kreck.
    p. cm. — (Graduate studies in mathematics ; v. 110)
  Includes bibliographical references and index.
  ISBN 978-0-8218-4898-2 (alk. paper)
  1. Algebraic topology. 2. Differential topology. I. Title.

QA612.K7 2010
514′.2—dc22
                                                              2009037982

Copying and reprinting. Individual readers of this publication, and nonprofit libraries acting for them, are permitted to make fair use of the material, such as to copy a chapter for use in teaching or research. Permission is granted to quote brief passages from this publication in reviews, provided the customary acknowledgment of the source is given.

Republication, systematic copying, or multiple reproduction of any material in this publication is permitted only under license from the American Mathematical Society. Requests for such permission should be addressed to the Acquisitions Department, American Mathematical Society, 201 Charles Street, Providence, Rhode Island 02904-2294 USA. Requests can also be made by e-mail to `reprint-permission@ams.org`.

© 2010 by the American Mathematical Society. All rights reserved.
The American Mathematical Society retains all rights
except those granted to the United States Government.
Printed in the United States of America.

∞ The paper used in this book is acid-free and falls within the guidelines
established to ensure permanence and durability.
Visit the AMS home page at `http://www.ams.org/`

10 9 8 7 6 5 4 3 2 1   15 14 13 12 11 10

Contents

INTRODUCTION	ix
Chapter 0. A quick introduction to stratifolds	1
Chapter 1. Smooth manifolds revisited	5
§1. A word about structures	5
§2. Differential spaces	6
§3. Smooth manifolds revisited	8
§4. Exercises	11
Chapter 2. Stratifolds	15
§1. Stratifolds	15
§2. Local retractions	18
§3. Examples	19
§4. Properties of smooth maps	25
§5. Consequences of Sard's Theorem	27
§6. Exercises	29
Chapter 3. Stratifolds with boundary: c-stratifolds	33
§1. Exercises	38
Chapter 4. $\mathbb{Z}/2$-homology	39
§1. Motivation of homology	39
§2. $\mathbb{Z}/2$-oriented stratifolds	41
§3. Regular stratifolds	43
§4. $\mathbb{Z}/2$-homology	45

§5.	Exercises	51

Chapter 5. The Mayer-Vietoris sequence and homology groups of spheres — 55

§1.	The Mayer-Vietoris sequence	55
§2.	Reduced homology groups and homology groups of spheres	61
§3.	Exercises	64

Chapter 6. Brouwer's fixed point theorem, separation, invariance of dimension — 67

§1.	Brouwer's fixed point theorem	67
§2.	A separation theorem	68
§3.	Invariance of dimension	69
§4.	Exercises	70

Chapter 7. Homology of some important spaces and the Euler characteristic — 71

§1.	The fundamental class	71
§2.	$\mathbb{Z}/2$-homology of projective spaces	72
§3.	Betti numbers and the Euler characteristic	74
§4.	Exercises	77

Chapter 8. Integral homology and the mapping degree — 79

§1.	Integral homology groups	79
§2.	The degree	83
§3.	Integral homology groups of projective spaces	86
§4.	A comparison between integral and $\mathbb{Z}/2$-homology	88
§5.	Exercises	89

Chapter 9. A comparison theorem for homology theories and CW-complexes — 93

§1.	The axioms of a homology theory	93
§2.	Comparison of homology theories	94
§3.	CW-complexes	98
§4.	Exercises	99

Chapter 10. Künneth's theorem — 103

§1.	The cross product	103
§2.	The Künneth theorem	106
§3.	Exercises	109

Contents   vii

Chapter 11. Some lens spaces and quaternionic generalizations	111
§1. Lens spaces	111
§2. Milnor's 7-dimensional manifolds	115
§3. Exercises	117
Chapter 12. Cohomology and Poincaré duality	119
§1. Cohomology groups	119
§2. Poincaré duality	121
§3. The Mayer-Vietoris sequence	123
§4. Exercises	125
Chapter 13. Induced maps and the cohomology axioms	127
§1. Transversality for stratifolds	127
§2. The induced maps	129
§3. The cohomology axioms	132
§4. Exercises	133
Chapter 14. Products in cohomology and the Kronecker pairing	135
§1. The cross product and the Künneth theorem	135
§2. The cup product	137
§3. The Kronecker pairing	141
§4. Exercises	145
Chapter 15. The signature	147
§1. Exercises	152
Chapter 16. The Euler class	153
§1. The Euler class	153
§2. Euler classes of some bundles	155
§3. The top Stiefel-Whitney class	159
§4. Exercises	159
Chapter 17. Chern classes and Stiefel-Whitney classes	161
§1. Exercises	165
Chapter 18. Pontrjagin classes and applications to bordism	167
§1. Pontrjagin classes	167
§2. Pontrjagin numbers	170
§3. Applications of Pontrjagin numbers to bordism	172
§4. Classification of some Milnor manifolds	174

§5.	Exercises	175

Chapter 19. Exotic 7-spheres — 177
  §1. The signature theorem and exotic 7-spheres — 177
  §2. The Milnor spheres are homeomorphic to the 7-sphere — 181
  §3. Exercises — 184

Chapter 20. Relation to ordinary singular (co)homology — 185
  §1. $SH_k(X)$ is isomorphic to $H_k(X;\mathbb{Z})$ for CW-complexes — 185
  §2. An example where $SH_k(X)$ and $H_k(X)$ are different — 187
  §3. $SH^k(M)$ is isomorphic to ordinary singular cohomology — 188
  §4. Exercises — 190

Appendix A. Constructions of stratifolds — 191
  §1. The product of two stratifolds — 191
  §2. Gluing along part of the boundary — 192
  §3. Proof of Proposition 4.1 — 194

Appendix B. The detailed proof of the Mayer-Vietoris sequence — 197

Appendix C. The tensor product — 209

Bibliography — 215

Index — 217

INTRODUCTION

In this book we present some basic concepts and results from algebraic and differential topology. We do this in the framework of differential topology. Homology groups of spaces are one of the central tools of algebraic topology. These are abelian groups associated to topological spaces which measure certain aspects of the complexity of a space.

The idea of homology was originally introduced by Poincaré in 1895 [**Po**] where homology classes were represented by certain global geometric objects like closed submanifolds. The way Poincaré introduced homology in this paper is the model for our approach. Since some basics of differential topology were not yet far enough developed, certain difficulties occurred with Poincaré's original approach. Three years later he overcame these difficulties by representing homology classes using sums of locally defined objects from combinatorics, in particular singular simplices, instead of global differential objects. The singular and simplicial approaches to homology have been very successful and up until now most books on algebraic topology follow them and related elaborations or variations.

Poincaré's original idea for homology came up again many years later, when in the 1950's Thom [**Th 1**] invented and computed the bordism groups of smooth manifolds. Following on from Thom, Conner and Floyd [**C-F**] introduced singular bordism as a generalized homology theory of spaces in the 1960's. This homology theory is much more complicated than ordinary homology, since the bordism groups associated to a point are complicated abelian groups, whereas for ordinary homology they are trivial except in degree 0. The easiest way to simplify the bordism groups of a point is to generalize manifolds in an appropriate way, such that in particular the cone over a closed manifold of dimension > 0 is such a generalized manifold. There are several approaches in the literature in this direction but they are at a more advanced level. We hope it is useful to present an approach to ordinary homology which reflects the spirit of Poincaré's original idea and is written as an introductory text. For another geometric approach to (co)homology see [**B-R-S**].

As indicated above, the key for passing from singular bordism to ordinary homology is to introduce generalized manifolds that are a certain kind of stratified space. These are topological spaces **S** together with a decomposition of **S** into manifolds of increasing dimension called the strata of **S**. There are many concepts of stratified spaces (for an important paper see [**Th 2**]), the most important examples being Whitney stratified spaces. (For a nice tour through the history of stratification theory and an alternative concept of smooth stratified spaces see [**Pf**].) We will introduce a new class of stratified spaces, which we call **stratifolds**. Here the decomposition of **S** into strata will be derived from another structure. We distinguish a certain algebra **C** of continuous functions which plays the role of smooth functions in the case of a smooth manifold. (For those familiar with the language of sheaves, **C** is the algebra of global sections of a subsheaf of the sheaf of continuous functions on **S**.) Others have considered such algebras before (see for example [**S-L**]), but we impose stronger conditions. More precisely, we use the language of differential spaces [**Si**] and impose on this additional conditions. The conditions we impose on the algebra **C** provide the decomposition of **S** into its strata, which are smooth manifolds.

It turns out that basic concepts from differential topology like Sard's theorem, partitions of unity and transversality generalize to stratifolds and this allows for a definition of homology groups based on stratifolds which we call "stratifold homology". For many spaces this agrees with the most common and most important homology groups: singular homology groups (see below). It is rather easy and intuitive to derive the basic properties of homology groups in the world of stratifolds. These properties allow computation of homology groups and straightforward constructions of important homology classes like the fundamental class of a closed smooth oriented manifold or, more generally, of a compact stratifold. We also define stratifold cohomology groups (but only for smooth manifolds) by following an idea of Quillen [**Q**], who gave a geometric construction of cobordism groups, the cohomology theory associated to singular bordism. Again, certain important cohomology classes occur very naturally in this description, in particular the characteristic classes of smooth vector bundles over smooth oriented manifolds. Another useful aspect of this approach is that one of the most fundamental results, namely Poincaré duality, is almost a triviality. On the other hand, we do not develop much homological algebra and so related features of homology are not covered: for example the general Künneth theorem and the universal coefficient theorem.

From (co)homology groups one can derive important invariants like the Euler characteristic and the signature. These invariants play a significant role in some of the most spectacular results in differential topology. As a highlight we present Milnor's exotic 7-spheres (using a result of Thom which we do not prove in this book).

We mentioned above that Poincaré left his original approach and defined homology in a combinatorial way. It is natural to ask whether the definition of stratifold homology in this book is equivalent to the usual definition of singular homology. Both constructions satisfy the Eilenberg-Steenrod axioms for a homology theory and so, for a large class of spaces including all spaces which are homotopy equivalent to CW-complexes, the theories are equivalent. There is also an axiomatic characterization of cohomology for smooth manifolds which implies that the stratifold cohomology groups of smooth manifolds are equivalent to their singular cohomology groups. We consider these questions in chapter 20. It was a surprise to the author to find out that for more general spaces than those which are homotopy equivalent to CW-complexes, our homology theory is different from ordinary singular homology. This difference occurs already for rather simple spaces like the one-point compactifications of smooth manifolds!

The previous paragraphs indicate what the main themes of this book will be. Readers should be familiar with the basic notions of point set topology and of differential topology. We would like to stress that one can start reading the book if one only knows the definition of a topological space and some basic examples and methods for creating topological spaces and concepts like Hausdorff spaces and compact spaces. From differential topology one only needs to know the definition of smooth manifolds and some basic examples and concepts like regular values and Sard's theorem. The author has given introductory courses on algebraic topology which start with the presentation of these prerequisites from point set and differential topology and then continue with chapter 1 of this book. Additional information like orientation of manifolds and vector bundles, and later on transversality, was explained was explained when it was needed. Thus the book can serve as a basis for a combined introduction to differential and algebraic topology.

It also allows for a quick presentation of (co)homology in a course about differential geometry.

As with most mathematical concepts, the concept of stratifolds needs some time to get used to. Some readers might want to see first what stratifolds are good for before they learn the details. For those readers I have collected a few basics about stratifolds in chapter 0. One can jump from there directly to chapter 4, where stratifold homology groups are constructed.

I presented the material in this book in courses at Mainz (around 1998) and Heidelberg Universities. I would like to thank the students and the assistants in these courses for their interest and suggestions for improvements. Thanks to Anna Grinberg for not only drawing the figures but also for careful reading of earlier versions and for several stimulating discussions. Also many thanks to Daniel Müllner and Martin Olbermann for their help. Diarmuid Crowley has read the text carefully and helped with the English (everything not appropriate left over falls into the responsibility of the author). Finally Peter Landweber read the final version and suggested improvements with a care I could never imagine. Many thanks to both of them. I had several fruitful discussions with Gerd Laures, Wilhelm Singhof, Stephan Stolz, and Peter Teichner about the fundamental concepts. Theodor Bröcker and Don Zagier have read a previous version of the book and suggested numerous improvements. The book was carefully refereed and I obtained from the referees valuable suggestions for improvements. I would like to thank these colleagues for their generous help. Finally, I would like to thank Dorothea Heukäufer and Ursula Jagtiani for the careful typing.

Chapter 0

A quick introduction to stratifolds

In this chapter we say as much as one needs to say about stratifolds in order to proceed directly to chapter 4 where homology with $\mathbb{Z}/2$-coefficients is constructed. We do it in a completely informal way that does not replace the definition of stratifolds. But some readers might want to see what stratifolds are good for before they study their definition and basic properties.

An n-dimensional **stratifold S** is a topological space **S** together with a class of distinguished continuous functions $f : \mathbf{S} \to \mathbb{R}$ called **smooth functions**. Stratifolds are generalizations of smooth manifolds M where the distinguished class of smooth functions are the C^∞-functions. The distinguished class of smooth functions on a stratifold **S** leads to a decomposition of **S** into disjoint smooth manifolds $\mathbf{S}^i$ of dimension i where $0 \leq i \leq n$, the dimension of **S**. We call the $\mathbf{S}^i$ the **strata** of **S**. An n-dimensional stratifold is a smooth manifold if and only if $\mathbf{S}^i = \varnothing$ for $i < n$.

To obtain a feeling for stratifolds we consider an important example. Let M be a smooth n-dimensional manifold. Then we consider the open cone over M

$$\overset{\circ}{C}M := M \times [0,1)/_{M \times \{0\}},$$

i.e., we consider the half open cylinder over M and collapse $M \times \{0\}$ to a point.

1

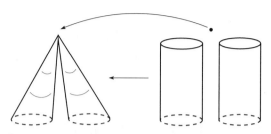

Now, we make $\overset{\circ}{C}M$ an $(n+1)$-dimensional stratifold by describing its distinguished class of smooth functions. These are the continuous functions
$$f : \overset{\circ}{C}M \to \mathbb{R},$$
such that $f|_{M\times(0,1)}$ is a smooth function on the smooth manifold $M \times (0,1)$ and there is an $\epsilon > 0$ such that $f|_{M\times[0,\epsilon)/M\times\{0\}}$ is constant. In other words, the function is locally constant near the cone point $M \times \{0\}/_{M\times\{0\}} \in \overset{\circ}{C}M$. The strata of this $(n+1)$-dimensional stratifold $\mathbf{S}$ turn out to be $\mathbf{S}^0 = M \times \{0\}/_{M\times\{0\}}$, the cone point, which is a 0-dimensional smooth manifold, $\mathbf{S}^i = \varnothing$ for $0 < i < n+1$ and $\mathbf{S}^{n+1} = M \times (0,1)$.

One can generalize this construction and make the open cone over any n-dimensional stratifold $\mathbf{S}$ an $(n+1)$-dimensional stratifold $\overset{\circ}{C}\mathbf{S}$. The strata of $\overset{\circ}{C}\mathbf{S}$ are: $(\overset{\circ}{C}\mathbf{S})^0 = \text{pt}$, the cone point, and for $1 \le i \le n+1$ we have $(\overset{\circ}{C}\mathbf{S})^i = \mathbf{S}^{i-1} \times (0,1)$, the open cylinder over the $(i-1)$-stratum of $\mathbf{S}$.

Stratifolds are defined so that most basic tools from differential topology for manifolds generalize to stratifolds.

- For each covering of a stratifold $\mathbf{S}$ one has a subordinate partition of unity consisting of smooth functions.
- One can define regular values of a smooth function $f : \mathbf{S} \to \mathbb{R}$ and show that if t is a regular value, then $f^{-1}(t)$ is a stratifold of dimension $n-1$ where the smooth functions of $f^{-1}(t)$ are simply the restrictions of the smooth functions of $\mathbf{S}$.
- Sard's theorem can be applied to show that the regular values of a smooth function $\mathbf{S} \to \mathbb{R}$ are a dense subset of $\mathbb{R}$.

As always, when we define mathematical objects like groups, vector spaces, manifolds, etc., we define the "allowed maps" between these objects, like homomorphisms, linear maps, smooth maps. In the case of stratifolds we do the same and call the "allowed maps" morphisms. A **morphism** $f : \mathbf{S} \to \mathbf{S}'$ is a continuous map $f : \mathbf{S} \to \mathbf{S}'$ such that for each smooth function $\rho : \mathbf{S}' \to \mathbb{R}$ the composition $\rho f : \mathbf{S} \to \mathbb{R}$ is a smooth function on $\mathbf{S}$. It is a nice exercise to show that the morphisms between smooth manifolds are precisely the smooth maps. A bijective map $f : \mathbf{S} \to \mathbf{S}'$ is called an **isomorphism** if f and f^{-1} are both morphisms. Thus in the case of smooth manifolds an isomorphism is the same as a diffeomorphism.

Next we consider stratifolds with boundary. For those who know what an n-dimensional manifold W with boundary is, it is clear that W is a topological space together with a distinguished closed subspace $\partial W \subseteq W$ such that $W - \partial W =: \overset{\circ}{W}$ is a n-dimensional smooth manifold and ∂W is a $(n-1)$-dimensional smooth manifold. For our purposes it is enough to imagine the same picture for stratifolds with boundary. An n-dimensional stratifold $\mathbf{T}$ with boundary is a topological space $\mathbf{T}$ together with a closed subspace $\partial \mathbf{T}$, the structure of a n-dimensional stratifold on $\overset{\circ}{\mathbf{T}} = \mathbf{T} - \partial \mathbf{T}$, the structure of an $(n-1)$-dimensional stratifold on $\partial \mathbf{T}$ and an additional structure (a collar) which we will not describe here. We call a stratifold with boundary a **c-stratifold** because of this collar.

The most important example of a smooth n-dimensional manifold with boundary is the half open cylinder $M \times [0,1)$ over a $(n-1)$-dimensional manifold M, where $\partial M = M \times \{0\}$. Similarly, if $\mathbf{S}$ is a stratifold, then we give $\mathbf{S} \times [0,1)$ the structure of a stratifold $\mathbf{T}$ with $\partial \mathbf{T} = \mathbf{S} \times \{0\}$. In the world of stratifolds the most important example of a c-stratifold is the closed cone over a smooth $(n-1)$-dimensional manifold M. This is denoted by

$$CM := M \times [0,1]/_{M \times \{0\}},$$

where $\partial CM := M \times \{1\}$. More generally, for an $(n-1)$-dimensional stratifold $\mathbf{S}$ one can give the closed cone

$$C\mathbf{S} := \mathbf{S} \times [0,1]/_{\mathbf{S} \times \{0\}}$$

the structure of a c-stratifold with $\partial C\mathbf{S} = \mathbf{S} \times \{1\}$.

If $\mathbf{T}$ and $\mathbf{T}'$ are stratifolds and $f : \partial \mathbf{T} \to \partial \mathbf{T}'$ is an isomorphism one can paste $\mathbf{T}$ and $\mathbf{T}'$ together via f. As a topological space one takes the disjoint union $\mathbf{T} \sqcup \mathbf{T}'$ and introduces the equivalence relation which identifies $x \in \partial \mathbf{T}$ with $f(x) \in \partial \mathbf{T}'$. There is a canonical way to give this space a stratifold structure. We denote the resulting stratifold by

$$\mathbf{T} \cup_f \mathbf{T}'.$$

If $\partial \mathbf{T} = \partial \mathbf{T}'$ and $f = \mathrm{id}$, the identity map, we write

$$\mathbf{T} \cup \mathbf{T}'$$

instead of $\mathbf{T} \cup_f \mathbf{T}'$.

Instead of gluing along the full boundary we can glue along some components of the boundary, as shown below.

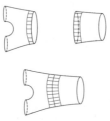

If a reader decides to jump from this chapter straight to homology (chapter 4), I recommend that he or she think of stratifolds as mathematical objects very similar to smooth manifolds, keeping in mind that in the world of stratifolds constructions like the cone over a manifold or even a stratifold are possible.

Chapter 1

Smooth manifolds revisited

Prerequisites: We assume that the reader is familiar with some basic notions from point set topology and differentiable manifolds. Actually rather little is needed for the beginning of this book. For example, it is sufficient to know [**Jä**, ch. 1 and 3] as background from point set topology. For the first chapters, all we need to know from differential topology is the definition of smooth ($= C^\infty$) manifolds (without boundary) and smooth ($= C^\infty$) maps (see for example [**Hi**, sec. I.1 and I.4]) or the corresponding chapters in [**B-J**]). In later chapters, where more background is required, the reader can find this in the cited literature.

1. A word about structures

Most definitions or concepts in modern mathematics are of the following type: a mathematical object is a set together with additional information called a structure. For example a group is a set G together with a map $G \times G \to G$, the multiplication, or a topological space is a set X together with certain subsets, the open subsets. Often the set is already equipped with a structure of one sort and one adds another structure, for example a vector space is an abelian group together with a second structure given by scalar multiplication, or a smooth manifold is a topological space together with a smooth atlas. Given such a structure one defines certain classes of "allowed" maps (often called morphisms) which respect this structure in a certain sense: for example group homomorphisms or continuous maps. The real numbers $\mathbb{R}$ admit many different structures: they are a group, a field, a vector space, a metric space, a topological space, a smooth manifold and so on. The "allowed" maps from a set with a structure to $\mathbb{R}$ with appropriate structure frequently play a leading role.

In this section we will define a structure on a topological space by specifying certain maps to the real numbers. This is done in such a way that the allowed maps are the maps specifying the structure. In other words, we give the allowed maps (morphisms) and in this way we define a structure. For example, we will define a smooth manifold M by specifying the C^∞-maps to $\mathbb{R}$. This stresses the role played by the allowed maps to $\mathbb{R}$ which are of central importance in many areas of mathematics, in particular analysis.

2. Differential spaces

We introduce the language of differential spaces [**Si**], which are topological spaces together with a distinguished set of continuous functions fulfilling certain properties. To formulate these properties the following notion is useful: if X is a topological space, we denote the set of continuous functions from X to $\mathbb{R}$ by $C^0(X)$.

Definition: *A subset* $\mathbf{C} \subset C^0(X)$ *is called an* **algebra** *of continuous functions if for $f, g \in \mathbf{C}$ the sum $f + g$, the product fg and all constant functions are in $\mathbf{C}$.*

The concept of an algebra, a vector space that at the same time is a ring fulfilling the obvious axioms, is more general, but here we only need algebras which are contained in $C^0(X)$.

For example, $C^0(X)$ itself is an algebra, and for that reason we call $\mathbf{C}$ a **subalgebra** of $C^0(X)$. The set of the constant functions is a subalgebra. If $U \subset \mathbb{R}^k$ is an open subset, we denote the set of functions $f : U \longrightarrow \mathbb{R}$, where all partial derivatives of all orders exist, by $C^\infty(U)$. This is a subalgebra in $C^0(U)$. More generally, if M is a k-dimensional smooth manifold then the set of smooth functions on M, denoted $C^\infty(M)$, is a subalgebra in $C^0(M)$.

Continuity is an example of a property of functions which can be decided locally, i.e., a function $f : X \longrightarrow \mathbb{R}$ is continuous if and only if for all $x \in X$ there is an open neighbourhood U of x such that $f|_U$ is continuous. The following is an equivalent—more complicated looking—formulation where we don't need to know what it means for $f|_U$ to be continuous. A function $f : X \to \mathbb{R}$ is continuous if and only if for each $x \in X$ there is an open neighbourhood U and a continuous function g such that $f|_U = g|_U$. Since this formulation makes sense for an arbitrary set of functions **C**, we define:

Definition: *Let **C** be a subalgebra of the algebra of continuous functions $f: X \to \mathbb{R}$. We say that **C** is **locally detectable** if a function $h: X \longrightarrow \mathbb{R}$ is contained in **C** if and only if for all $x \in X$ there is an open neighbourhood U of x and $g \in$ **C** such that $h|_U = g|_U$.*

As mentioned above, the set of continuous functions $C^0(X)$ is locally detectable. Similarly, if M is a smooth manifold, then $C^\infty(M)$ is locally detectable.

For those familiar with the language of sheaves it is obvious that $(X, \mathbf{C})$ is equivalent to a topological space X together with a **subsheaf of the sheaf of continuous functions**. If such a subsheaf is given, the global sections give a subalgebra **C** of $C^0(X)$, which by the properties of a sheaf is locally detectable. In turn, if a locally detectable subalgebra $\mathbf{C} \subset C^0(X)$ is given, then for an open subset U of X we define $\mathbf{C}(U)$ as the functions $f: U \to \mathbb{R}$ such that for each $x \in U$ there is an open neighbourhood V and $g \in \mathbf{C}$ with $g|_V = f|_V$. Since **C** is locally detectable, this gives a presheaf, whose associated sheaf is the sheaf corresponding to **C**.

We can now define differential spaces.

Definition: *A **differential space** is a pair $(X, \mathbf{C})$, where X is a topological space and $\mathbf{C} \subset C^0(X)$ is a locally detectable subalgebra of the algebra of continuous functions (or equivalently a space X together with a subsheaf of the sheaf of continuous functions on X) satisfying the condition:*

For all $f_1, \ldots, f_k \in \mathbf{C}$ and smooth functions $g: \mathbb{R}^k \longrightarrow \mathbb{R}$, the function

$$x \mapsto g(f_1(x), \ldots, f_k(x))$$

*is in **C**.*

This condition is clearly desirable in order to construct new elements of **C** by composition with smooth maps and it holds for smooth manifolds by the chain rule. In particular k-dimensional smooth manifolds are differential spaces and this is the fundamental class of examples which will be the model for our generalization to stratifolds in the next chapter.

From a differential space $(X, \mathbf{C})$, one can construct new differential spaces. For example, if $Y \subset X$ is a subspace, we define $\mathbf{C}(Y)$ to contain those functions $f : Y \longrightarrow \mathbb{R}$ such that for all $x \in Y$, there is a $g : X \longrightarrow \mathbb{R}$ in $\mathbf{C}$ such that $f|_V = g|_V$ for some open neighbourhood V of x in Y. The reader should check that $(Y, \mathbf{C}(Y))$ is a differential space.

There is another algebra associated to a subspace Y in X, namely the restriction of all elements in $\mathbf{C}$ to Y. Later we will consider differential spaces with additional properties which guarantee that $\mathbf{C}(Y)$ is equal to the restriction of elements in $\mathbf{C}$ to Y, if Y is a closed subspace.

For the generalization to stratifolds it is useful to note that one can define smooth manifolds in the language of differential spaces. To prepare for this, we need a way to compare differential spaces.

Definition: *Let $(X, \mathbf{C})$ and $(X', \mathbf{C}')$ be differential spaces. A homeomorphism $f : X \longrightarrow X'$ is called an* **isomorphism** *if for each $g \in \mathbf{C}'$ and $h \in \mathbf{C}$, we have $gf \in \mathbf{C}$ and $hf^{-1} \in \mathbf{C}'$.*

The slogan is: composition with f stays in $\mathbf{C}$ and with f^{-1} stays in $\mathbf{C}'$. Obviously the identity map is an isomorphism from $(X, \mathbf{C})$ to $(X, \mathbf{C})$. If $f : X \to X'$ and $f' : X' \to X''$ are isomorphisms then $f'f : X \to X''$ is an isomorphism. If f is an isomorphism then f^{-1} is an isomorphism.

For example, if X and X' are open subspaces of $\mathbb{R}^k$ equipped with the algebra of smooth functions, then an isomorphism f is the same as a diffeomorphism from X to X': a bijective map such that the map and its inverse are smooth ($= C^\infty$) maps. This equivalence is due to the fact that a map g from an open subset U of $\mathbb{R}^k$ to an open subset V of $\mathbb{R}^n$ is smooth if and only if all coordinate functions are smooth. (For a similar discussion, see the end of this chapter.)

3. Smooth manifolds revisited

We recall that if $(X, \mathbf{C})$ is a differential space and U an open subspace, the algebra $\mathbf{C}(U)$ is defined as the continuous maps $f : U \to \mathbb{R}$ such that for each $x \in U$ there is an open neighbourhood $V \subset U$ of x and $g \in \mathbf{C}$ such that $g|_V = f|_V$. We remind the reader that $(U, \mathbf{C}(U))$ is a differential space.

3. Smooth manifolds revisited

Definition: *A k-dimensional* **smooth manifold** *is a differential space $(M, \mathbf{C})$ where M is a Hausdorff space with a countable basis of its topology, such that for each $x \in M$ there is an open neighbourhood $U \subseteq M$, an open subset $V \subset \mathbb{R}^k$ and an isomorphism*

$$\varphi : (V, C^\infty(V)) \to (U, \mathbf{C}(U)).$$

The slogan is: a k-dimensional smooth manifold is a differential space which is locally isomorphic to $\mathbb{R}^k$.

To justify this definition of this well known mathematical object, we have to show that it is equivalent to the definition based on a maximal smooth atlas. Starting from the definition above, we consider all isomorphisms $\varphi : (V, C^\infty(V)) \to (U, \mathbf{C}(U))$ from the definition above and note that their coordinate changes $\varphi^{-1}\varphi' : (\varphi')^{-1}(U \cap U') \to \varphi^{-1}(U \cap U')$ are smooth maps and so the maps $\varphi : V \to U$ give a maximal smooth atlas on M. In turn if a smooth atlas $\varphi : V \to U \subset M$ is given, then we define $\mathbf{C}$ as the continuous functions $f : M \to \mathbb{R}$ such that for all φ in the smooth atlas $f\varphi : V \to \mathbb{R}$ is in $C^\infty(V)$.

We want to introduce the important concept of the germ of a function. Let C be a set of functions from X to $\mathbb{R}$, and let $x \in X$. We define an equivalence relation on C by setting f equivalent to g if and only if there is an open neighbourhood V of x such that $f|_V = g|_V$. We call the equivalence class represented by f the **germ** of f at x and denote this equivalence class by $[f]_x$. We denote the set of germs of functions at x by C_x. This definition of germs is different from the standard one which only considers equivalence classes of functions defined on some open neighbourhood of x. For differential spaces these sets of equivalence classes are the same, since if $f : U \to \mathbb{R}$ is defined on some open neighbourhood of x, then there is a $g \in \mathbf{C}$ such that on some smaller neighbourhood V we have $f|_V = g|_V$.

To prepare for the definition of stratifolds in the next chapter, we recall the definition of the tangent space at a point $x \in M$ in terms of derivations. Let $(X, \mathbf{C})$ be a differential space. For a point $x \in X$, we consider the germs of functions at x, $\mathbf{C}_x$. If $f \in \mathbf{C}$ and $g \in \mathbf{C}$ are representatives of germs at x, then the sum $f + g$ and the product $f \cdot g$ represent well-defined germs denoted $[f]_x + [g]_x \in \mathbf{C}_x$ and $[f]_x \cdot [g]_x \in \mathbf{C}_x$.

Definition: Let $(X, \mathbf{C})$ be a differential space. A **derivation** at $x \in X$ is a map from the germs of functions at x

$$\alpha : \mathbf{C}_x \longrightarrow \mathbb{R}$$

such that

$$\alpha([f]_x + [g]_x) = \alpha([f]_x) + \alpha([g]_x),$$

$$\alpha([f]_x \cdot [g]_x) = \alpha([f]_x) \cdot g(x) + f(x) \cdot \alpha([g]_x),$$

and

$$\alpha([c]_x \cdot [f]_x) = c \cdot \alpha([f]_x)$$

for all $f, g \in \mathbf{C}$ and $[c]_x$ the germ of the constant function which maps all $y \in X$ to $c \in \mathbb{R}$.

If $U \subset \mathbb{R}^k$ is an open set and $v \in \mathbb{R}^k$, the Leibniz rule says that for $x \in U$, the map

$$\alpha_v : C^\infty(U)_x \longrightarrow \mathbb{R}$$

$$[f]_x \longmapsto df_x(v)$$

is a derivation. Thus the derivative in the direction of v is a derivation which justifies the name.

If α and β are derivations, then $\alpha + \beta$ mapping $[f]_x$ to $\alpha([f]_x) + \beta([f]_x)$ is a derivation, and if $t \in \mathbb{R}$ then $t\alpha$ mapping $[f]_x$ to $t\alpha([f]_x)$ is a derivation. Thus the derivations at $x \in X$ form a vector space.

Definition: Let $(X, \mathbf{C})$ be a differential space and $x \in X$. The vector space of derivations at x is called the **tangent space** of X at x and denoted by $T_x X$.

This notation is justified by the fact that if M is a k-dimensional smooth manifold, which we interpret as a differential space $(M, C^\infty(M))$, then the definition above is one of the equivalent definitions of the tangent space [**B-J**, p. 14]. The isomorphism is given by the map above associating to a tangent vector v at x the derivation which maps f to $df_x(v)$. In particular, $\dim T_x X = k$.

We have already defined isomorphisms between differential spaces. We also want to introduce morphisms. If the differential spaces are smooth manifolds, then the morphisms will be the smooth maps. To generalize the definition of smooth maps to differential spaces, we reformulate the definition of smooth maps between manifolds.

If M is an m-dimensional smooth manifold and U is an open subset of $\mathbb{R}^k$ then a map $f : M \longrightarrow U$ is a smooth map if and only if all components $f_i : M \longrightarrow \mathbb{R}$ are in $C^\infty(M)$ for $1 \leq i \leq k$. If we don't want to use components we can equivalently say that f is smooth if and only if for all $\rho \in C^\infty(U)$ we have $\rho f \in C^\infty(M)$. This is the logic behind the following definition. Let $(X, \mathbf{C})$ be a differential space and $(X', \mathbf{C}')$ another differential space. Then we define a **morphism** f from $(X, \mathbf{C})$ to $(X', \mathbf{C}')$ as a continuous map $f : X \longrightarrow X'$ such that for all $\rho \in \mathbf{C}'$ we have $\rho f \in \mathbf{C}$. We denote the set of morphisms by $\mathbf{C}(X, X')$. The following properties are obvious from the definition:

(1) $\text{id} : (X, \mathbf{C}) \longrightarrow (X, \mathbf{C})$ is a morphism,
(2) if $f : (X, \mathbf{C}) \longrightarrow (X', \mathbf{C}')$ and $g : (X', \mathbf{C}') \longrightarrow (X'', \mathbf{C}'')$ are morphisms, then $gf : (X, \mathbf{C}) \longrightarrow (X'', \mathbf{C}'')$ is a morphism,
(3) all elements of $\mathbf{C}$ are morphisms from X to $\mathbb{R}$,
(4) the isomorphisms (as defined above) are the morphisms
$$f : (X, \mathbf{C}) \longrightarrow (X', \mathbf{C}')$$
such that there is a morphism $g : (X', \mathbf{C}') \longrightarrow (X, \mathbf{C})$ with $gf = \text{id}_X$ and $fg = \text{id}_{X'}$.

We define the differential of a morphism as follows.

Definition: *Let $f : (X, \mathbf{C}) \to (X', \mathbf{C}')$ be a morphism. Then for each $x \in X$ the* **differential**
$$df_x : T_x X \to T_{f(x)} X'$$
is the map which sends a derivation α to α' where α' assigns to $[g]_{f(x)} \in \mathbf{C}'_{x'}$ the value $\alpha([gf]_x)$.

4. Exercises

(1) Let $U \subseteq \mathbb{R}^n$ be an open subset. Show that $(U, C^\infty(U))$ is equal to $(U, C(U))$ where the latter is the induced differential space structure which was described in this chapter.
(2) Give an example of a differential space $(X, C(X))$ and a subspace $Y \subseteq X$ such that the restriction of all functions in $C(X)$ to Y doesn't give a differential space structure.
(3) Let $(X, C(X))$ be a differential space and $Z \subseteq Y \subseteq X$ be two subspaces. We can give Z two We can give Z two differential structures:

First by inducing the structure from $(X, C(X))$ and the other one by first inducing the structure from $(X, C(X))$ to Y and then to Z. Show that both structures agree.

(4) We have associated to each smooth manifold with a maximal atlas a differential space which we called a smooth manifold and vice versa. Show that these associations are well defined and are inverse to each other.

(5) a) Let X be a topological space such that $X = X_1 \cup X_2$, a union of two open sets. Let $(X_1, C(X_1))$ and $(X_2, C(X_2))$ be two differential spaces which induce the same differential structure on $U = X_1 \cap X_2$. Give a differential structure on X which induces the differential structures on X_1 and on X_2.
b) Show that if both $(X_1, C(X_1))$ and $(X_2, C(X_2))$ are smooth manifolds and X is Hausdorff then X with this differential structure is a smooth manifold as well. Do we need to assume that X is Hausdorff or it is enough to assume that for both X_1 and X_2?

(6) Let $(M, C(M))$ be a smooth manifold and $U \subset M$ an open subset. Prove that $(U, C(U))$ is a smooth manifold.

(7) Prove or give a counterexample: Let $(X, C(X))$ be a differential space such that for every point $x \in X$ the dimension of the tangent space is equal to n, then it is a smooth manifold of dimension n.

(8) Show that the following differential spaces give the standard structure of a manifold on the following spaces:
a) S^n with the restriction of all smooth maps $f : \mathbb{R}^{n+1} \to \mathbb{R}$.
b) $\mathbb{RP}^n$ with all maps $f : \mathbb{RP}^n \to \mathbb{R}$ such that their composition with the quotient map $\pi : S^n \to \mathbb{RP}^n$ is smooth.
c) More generally, let M be a smooth manifold and let G be a finite group. Assume we have a smooth free action of G on M. Give a differential structure on the quotient space M/G which is a smooth manifold and such that the quotient map is a local diffeomorphism.

(9) Let $(M, C(M))$ be a smooth manifold and N be a closed submanifold of M. Show that the natural structure on N is given by the restrictions of the smooth maps $f : M \to \mathbb{R}$.

(10) Consider (S^n, C) where S^n is the n-sphere and C is the set of smooth functions which are locally constant near $(1, 0, 0, \ldots, 0)$. Show that (S^n, C) is a differential space but not a smooth manifold.

(11) Let $(X, C(X))$ and $(Y, C(Y))$ be two differential spaces and $f : X \to Y$ a morphism. For a point $x \in X$:
a) Show that composition induces a well-defined linear map $C_{f(x)} \to C_x$ between the germs.

4. Exercises

b) What can you say about the differential map if the above map is injective, surjective or an isomorphism?

(12) Show that the vector space of all germs of smooth functions at a point x in $\mathbb{R}^n$ is not finite dimensional for $n \geq 1$.

(13) Let M, N be two smooth manifolds and $f : M \to N$ a map. Show that f is smooth if and only if for every smooth map $g : N \to \mathbb{R}$ the composition is smooth.

(14) Show that the 2-torus $S^1 \times S^1$ is homeomorphic to the square $I \times I$ with opposite sides identified.

(15) a) Let M_1 and M_2 be connected n-dimensional manifolds and let $\phi_i : B^n \to M_i$ be two embeddings. Remove $\phi_i(\frac{1}{2}B^n)$ and for each $x \in \frac{1}{2}S_i^{n-1}$ identify in the disjoint union the points $\phi_1(x)$ and $\phi_2(x)$. Prove that this is a connected n-dimensional topological manifold.
b) If M_i are smooth manifolds and ϕ_i are smooth embeddings show that there is a smooth structure on this manifold which outside $M_i \setminus \phi_i(\frac{1}{2}D^n)$ agrees with the given smooth structures.

One can show that in the smooth case the resulting manifold is unique up to diffeomorphism. It is called the **connected sum**, denoted by $M_1 \# M_2$ and does not depend on the maps ϕ_i.

One can also show that every compact orientable surface is diffeomorphic to S^2 or a connected sum of tori $T^2 = S^1 \times S^1$ and every compact non-orientable surface is diffeomorphic to a connected sum of projective planes $\mathbb{RP}^2$.

Chapter 2

Stratifolds

Prerequisites: The main new ingredient is Sard's Theorem (see for example [B-J, chapt. 6] or [Hi, chapt. 3.1]). It is enough to know the statement of this important result.

1. Stratifolds

We will define a stratifold as a differential space with certain properties. The main feature of these properties is that there is a natural decomposition of a stratifold into subspaces which are smooth manifolds. We begin with the definition of this decomposition for a differential space.

Let $(\mathbf{S}, \mathbf{C})$ be a differential space. We define the subspace $\mathbf{S}^i := \{x \in \mathbf{S} \mid \dim T_x \mathbf{S} = i\}$.

By construction $\mathbf{S} = \bigsqcup_i \mathbf{S}^i$, i.e., $\mathbf{S}$ is the disjoint union (topological sum) of the subsets $\mathbf{S}^i$. We shall assume that the dimension of $T_x \mathbf{S}$ is finite for all points $x \in \mathbf{S}$. In chapter 1 we introduced the differential spaces $(\mathbf{S}^i, \mathbf{C}(\mathbf{S}^i))$ given by the subspace $\mathbf{S}^i$ together with the induced algebra. Our first condition is that this differential space is a smooth manifold.

1a) We require that $(\mathbf{S}^i, \mathbf{C}(\mathbf{S}^i))$ is a smooth manifold (as defined in chapter 1).

Once this condition is fulfilled we write $C^\infty(\mathbf{S}^i)$ instead of $\mathbf{C}(\mathbf{S}^i)$. This smooth structure on $\mathbf{S}^i$ has the property that any smooth function can locally be extended to an element of $\mathbf{C}$. We want to strengthen this property by requiring that in a certain sense such an extension is unique. To formulate this we note that for points $x \in \mathbf{S}^i$, we have two sorts of germs of functions, namely $\mathbf{C}_x$, the germs of functions near x on $\mathbf{S}$, and $C^\infty(\mathbf{S}^i)_x$, the germs of smooth functions near x on $\mathbf{S}^i$, and our second condition requires that these sets of germs are equal. More precisely, condition 1b is as follows.

1b) Restriction defines for all $x \in \mathbf{S}^i$, a bijection
$$\mathbf{C}_x \xrightarrow{\cong} C^\infty(\mathbf{S}^i)_x$$
$$[f]_x \longrightarrow [f|_{\mathbf{S}^i}]_x.$$

Here the only new input is the injectivity, since the surjectivity follows from the definition. As a consequence, the tangent space of $\mathbf{S}$ at x is isomorphic to the tangent space of $\mathbf{S}^i$ at x. In particular we conclude that
$$\dim \mathbf{S}^i = i.$$

Conditions 1a and 1b give the most important properties of a stratifold. In addition we impose some other conditions which are common in similar contexts. To formulate them we introduce the following terminology and notation.

We call $\mathbf{S}^i$ the i-**stratum** of $\mathbf{S}$. In other concepts of spaces which are decomposed as smooth manifolds, the connected components of $\mathbf{S}^i$ are called the strata but we prefer to collect the i-dimensional strata into a single stratum. We call $\bigcup_{i \leq r} \mathbf{S}^i =: \Sigma^r$ the r-**skeleton** of $\mathbf{S}$.

Definition: *A k-dimensional* **stratifold** *is a differential space* $(\mathbf{S}, \mathbf{C})$, *where $\mathbf{S}$ is a locally compact (meaning each point is contained in a compact neighbourhood) Hausdorff space with countable basis, and the skeleta Σ^i are closed subspaces. In addition we assume:*

(1) *the conditions 1a and 1b are fulfilled, i.e., restriction gives a smooth structure on $\mathbf{S}^i$ and for each $x \in \mathbf{S}^i$ restriction gives an isomorphism*
$$i^* : \mathbf{C}_x \xrightarrow{\cong} C^\infty(\mathbf{S}^i)_x,$$

(2) $\dim T_x \mathbf{S} \leq k$ *for all $x \in \mathbf{S}$, i.e., all tangent spaces have dimension $\leq k$,*

(3) *for each $x \in \mathbf{S}$ and open neighbourhood $U \subset \mathbf{S}$ there is a nonnegative function $\rho \in \mathbf{C}$ such that $\rho(x) \neq 0$ and $\mathrm{supp}\, \rho \subseteq U$ (such a function is called a* **bump function***).*

We recall that the **support of a function** $f : X \to \mathbb{R}$ is $\operatorname{supp} f := \overline{\{x \mid f(x) \neq 0\}}$, the closure of the points where f is non-zero.

In our definition of a stratifold, the dimension k is always a finite number. One could easily define infinite dimensional stratifolds where the only difference is that in condition 2, we would require that $\dim T_x \mathbf{S}$ is finite for all $x \in \mathbf{S}$. Infinite dimensional stratifolds will play no role in this book.

Let us comment on these conditions. The most important conditions are 1a and 1b, which we have already explained previously. In particular, we recall that the smooth structure on $\mathbf{S}^i$ is determined by $\mathbf{C}$ which gives us a **stratification** of $\mathbf{S}$, a decomposition into smooth manifolds $\mathbf{S}^i$ of dimension i. The second condition says that the dimension of all non-empty strata is less than or equal to k. We don't assume that $\mathbf{S}^k \neq \varnothing$ which, at first glance, might look strange, but even in the definition of a k-dimensional manifold M, it is not required that $M \neq \varnothing$.

The third condition will be used later to show the existence of a partition of unity, an important tool to construct elements of $\mathbf{C}$. To do this, we will also use the topological conditions that the space is locally compact, Hausdorff, and has a countable basis. The other topological conditions on the skeleta and strata are common in similar contexts. For example, they guarantee that the **top stratum** $\mathbf{S}^k$ is open in $\mathbf{S}$, a useful and natural property. Here we note that the requirement that the skeleta are closed is equivalent to the requirement that for each $j > i$ we have $\overline{\mathbf{S}^i} \cap \mathbf{S}^j = \varnothing$. This topological condition roughly says that if we "walk" in $\mathbf{S}^i$ to a limit point outside $\mathbf{S}^i$, then this point sits in $\mathbf{S}^r$ for $r < i$. These conditions are common in similar contexts such as CW-complexes.

We have chosen the notation Σ^j for the j-skeleton since Σ^{k-1} is the singular set of $\mathbf{S}$ in the sense that $\mathbf{S} - \Sigma^{k-1} = \mathbf{S}^k$ is a smooth k-dimensional manifold. Thus if $\Sigma^{k-1} = \varnothing$, then $\mathbf{S}$ is a smooth manifold.

We call our objects stratifolds because on the one hand they are stratified spaces, while on the other hand they are in a certain sense very close to smooth manifolds even though stratifolds are much more general than smooth manifolds. As we will see, many of the fundamental tools of differential topology are available for stratifolds. In this respect smooth manifolds and stratifolds are not very different and deserve a similar name.

Remark: It's a nice property of smooth manifolds that once an algebra $\mathbf{C} \subset C^0(M)$ is given for a locally compact Hausdorff space M with countable basis, the question, whether $(M, \mathbf{C})$ is a smooth manifold is a local question. The same is true for stratifolds, since the conditions 1 – 3 are again local.

2. Local retractions

To obtain a better feeling for the central condition 1b, we give an alternative description. If $(\mathbf{S}, \mathbf{C})$ is a stratifold and $x \in \mathbf{S}^i$, we will construct an open neighborhood U_x of x in $\mathbf{S}$ and a morphism $r_x : U_x \to U_x \cap \mathbf{S}^i$ such that $r_x|_{U_x \cap \mathbf{S}^i} = \mathrm{id}_{U_x \cap \mathbf{S}^i}$. (Here we consider U_x as a differential space with the induced structure on an open subset as described in chapter 1.) Such a map is called a **local retraction** from U_x to $V_x := U_x \cap \mathbf{S}^i$. If one has a local retraction $r : U_x \to U_x \cap \mathbf{S}^i =: V_x$, we can use it to extend a smooth map $g : V_x \to \mathbb{R}$ to a map on U_x by gr. Thus composition with r gives a map

$$C^\infty(\mathbf{S}^i)_x \longrightarrow \mathbf{C}_x$$

mapping $[h]$ to $[hr]$, where we represent h by a map whose domain is contained in V_x. This gives an inverse of the isomorphism in condition 1b given by restriction.

To construct a retraction we choose an open neighborhood W of x in $\mathbf{S}^i$ such that W is the domain of a chart $\varphi : W \longrightarrow \mathbb{R}^i$ (we want that $\mathrm{im}\, \varphi = \mathbb{R}^i$ and we achieve this by starting with an arbitrary chart, which contains one whose image is an open ball which we identify with $\mathbb{R}^i$ by an appropriate diffeomorphism). Now we consider the coordinate functions $\varphi_j : W \longrightarrow \mathbb{R}$ of φ and consider for each $x \in W$ the germ represented by φ_j. By condition 1b there is an open neighbourhood $W_{j,x}$ of x in $\mathbf{S}$ and an extension $\hat\varphi_{j,x}$ of $\varphi_j|_{W_{j,x} \cap \mathbf{S}^i}$. We denote the intersection $\bigcap_{j=1}^i W_{j,x}$ by W_x and obtain a morphism $\hat\varphi_x : W_x \longrightarrow \mathbb{R}^i$ such that $y \mapsto (\hat\varphi_{1,x}(y), \ldots, \hat\varphi_{i,x}(y))$. For $y \in W_x \cap \mathbf{S}^i$ we have $\hat\varphi_x(y) = \varphi(y)$. Next we define

$$r : W_x \longrightarrow \mathbf{S}^i$$

$$z \mapsto \varphi^{-1}\hat\varphi_x(z).$$

For $y \in W_x \cap \mathbf{S}^i$ we have $r(y) = y$. Finally we define $U_x := r^{-1}(W_x \cap \mathbf{S}^i)$ and

$$r_x := r|_{U_x} : U_x \to U_x \cap \mathbf{S}^i = W_x \cap \mathbf{S}^i =: V_x$$

is the desired retraction.

We summarize these considerations.

Proposition 2.1. *(Local retractions) Let* $(\mathbf{S}, \mathbf{C})$ *be a stratifold. Then for* $x \in \mathbf{S}^i$ *there is an open neighborhood* U *of* x *in* $\mathbf{S}$*, an open neighbourhood* V *of* x *in* $\mathbf{S}^i$ *and a morphism*

$$r : U \to V$$

such that $U \cap \mathbf{S}^i = V$ *and* $r|_V = \mathrm{id}$. *Such a morphism is called a* **local retraction** *near* x.

If $r : U \to V$ *is a local retraction near* x, *then* r *induces an isomorphism*

$$C^\infty(\mathbf{S}^i)_x \to \mathbf{C}_x,$$

$$[h] \mapsto [hr],$$

the inverse of $i^* : \mathbf{C}_x \to C^\infty(\mathbf{S}^i)_x$.

The germ of local retractions near x *is unique, i.e., if* $r' : U' \to V'$ *is another local retraction near* x, *then there is a* $U'' \subset U \cap U'$ *such that* $r|_{U''} = r'|_{U''}$.

Note that one can use the local retractions to characterize elements of $\mathbf{C}$, namely a continuous function $f : \mathbf{S} \to \mathbb{R}$ is in $\mathbf{C}$ if and only if its restriction to all strata is smooth and it commutes with appropriate local retractions. This implies that if $f : \mathbf{S} \to \mathbb{R}$ is a nowhere zero morphism then $1/f$ is in $\mathbf{C}$.

3. Examples

The first class of examples is given by the smooth k-dimensional manifolds. These are the k-dimensional stratifolds with $\mathbf{S}^i = \varnothing$ for $i < k$. It is clear that such a stratifold gives a smooth manifold and in turn a k-dimensional manifold gives a stratifold. All conditions are obvious (for the existence of a bump function see [**B-J**, p. 66], or [**Hi**, p. 41].

Example 1: The most fundamental non-manifold example is the cone over a manifold. We define the open cone over a topological space Y as $Y \times [0, 1)/(Y \times \{0\}) =: \overset{\circ}{C}Y$. (We call it the open cone and use the notation $\overset{\circ}{C}Y$ to distinguish it from the (closed) cone $CY := Y \times [0, 1]/(Y \times \{0\})$.) We call the point $Y \times \{0\}/_{Y \times \{0\}}$ the top of the cone and abbreviate this as pt.

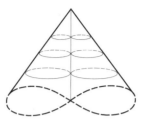

Let M be a k-dimensional compact smooth manifold. We consider the **open cone** over M and define an algebra making it a stratifold. We define the algebra $\mathbf{C} \subset C^0(\overset{\circ}{C}M)$ consisting of all functions in $C^0(\overset{\circ}{C}M)$ which are constant on some open neighbourhood U of the top of the cone and whose restriction to $M \times (0,1)$ is in $C^\infty(M \times (0,1))$. We want to show that $(\overset{\circ}{C}M, \mathbf{C})$ is a $(k+1)$-dimensional stratifold. It is clear from the definition of $\mathbf{C}$ that $\mathbf{C}$ is a locally detectable algebra, and that the condition in the definition of differential spaces is fulfilled.

So far we have seen that the open cone $(\overset{\circ}{C}M, \mathbf{C})$ is a differential space. We now check that the conditions of a stratifold are satisfied. Obviously, $\overset{\circ}{C}M$ is a Hausdorff space with a countable basis and, since M is compact, $\overset{\circ}{C}M$ is locally compact. The other topological properties of a stratifold are clear. We continue with the description of the stratification. For $x \neq \text{pt}$, the top of the cone, $\mathbf{C}_x$ is the set of germs of smooth functions on $M \times (0,1)$ near x, $T_x(\overset{\circ}{C}M) = T_x(M \times (0,1))$ which implies that $\dim T_x(\overset{\circ}{C}M) = k+1$. For $x = \text{pt}$, the top of the cone, $\mathbf{C}_x$ consists of simply the germs of constant functions. For the constant function 1 mapping all points to 1, we see for each derivation α, we have $\alpha(1) = \alpha(1 \cdot 1) = \alpha(1) \cdot 1 + 1 \cdot \alpha(1)$ implying $\alpha(1) = 0$. But since $\alpha([c] \cdot 1) = c \cdot \alpha(1)$, we conclude that $T_{\text{pt}}(\overset{\circ}{C}M) = 0$ and $\dim T_{\text{pt}}(\overset{\circ}{C}M) = 0$. Thus we have two non-empty strata: $M \times (0,1)$ and the top of the cone.

The conditions 1 and 2 are obviously fulfilled. It remains to show the existence of bump functions. Near points $x \neq \text{pt}$ the existence of a bump function follows from the existence of a bump function in $M \times (0,1)$ which we extend by 0 to $\overset{\circ}{C}M$. Near pt we first note that any open neighbourhood of pt contains an open neighbourhood of the form $M \times [0, \epsilon)/(M \times \{0\})$ for an appropriate $\epsilon > 0$. Then we choose a smooth function $\eta: [0,1) \to [0, \infty)$ which is 1 near 0 and 0 for $t \geq \epsilon$ (for the construction of such a function see

[**B-J**, p. 65]). With the help of η, we can now define the bump function

$$\rho([x,t]) := \eta(t)$$

which completes the proof that $(\overset{\circ}{C}M, \mathbf{C})$ is a $(k+1)$-dimensional stratifold. It has two non-empty strata: $\mathbf{S}^{k+1} = M \times (0,1)$ and $\mathbf{S}^0 = \text{pt}$.

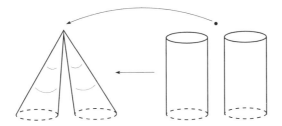

Remark: One might wonder if every smooth function on a stratum extends to a smooth function on $\mathbf{S}$. This is not the case as one can see from the open cone $\overset{\circ}{C}M$, where M is a compact non-empty smooth manifold. A smooth function on the top stratum $M \times (0,1)$ can be extended to the open cone if and only if it is constant on $M \times (0, \epsilon)$ for an appropriate $\epsilon > 0$.

Example 2: Let M be a non-compact m-dimensional manifold. The **one-point compactification** of M is the space M^+ consisting of M and an additional point $+$. The topology is given by defining open sets as the open sets of M together with the complements of compact subsets of M. The latter give the open neighbourhoods of $+$. It is easy to show that the one-point compactification is a Hausdorff space and has a countable basis. (For more information see e.g. [**Sch**].)

On M^+, we define the algebra $\mathbf{C}$ as the continuous functions which are constant on some open neighbourhood of $+$ and smooth on M. Then $(M^+, \mathbf{C})$ is an m-dimensional stratifold. All conditions except 3 are obvious. For the existence of a bump function near $+$ (near all other points use a bump function of M and extend it by 0 to $+$), let U be an open neighbourhood of $+$. By definition of the topology, $M - U =: A$ is a compact subset of M. Then one constructs another compact subset $B \subset M$ with $A \subset \overset{\circ}{B}$ (how?), and, starting from B instead of A, a third compact subset $C \subset M$ with $B \subset \overset{\circ}{C}$. Then B and $M - \overset{\circ}{C}$ are disjoint closed subsets of M and there is a smooth function $\rho : M \to (0, \infty)$ such that $\rho|_B = 0$ and $\rho|_{M-\overset{\circ}{C}} = 1$. We extend ρ to M^+ by mapping $+$ to 1 to obtain a bump function on M near $+$.

Thus we have given the one-point compactification of a smooth non-compact m-dimensional manifold M the structure of a stratifold $\mathbf{S} = M^+$, with non-empty strata $\mathbf{S}^m = M$ and $\mathbf{S}^0 = +$.

Example 3: The most natural examples of manifolds with singularities occur in algebraic geometry as algebraic varieties, i.e., zero sets of a family of polynomials. There is a natural but not completely easy way (and for that reason we don't give any details and refer to the thesis of Anna Grinberg [**G**]) to impose the structure of a stratifold on an algebraic variety (this proceeds in two steps, namely, one first shows that a variety is a Whitney stratified space and then one uses the retractions constructed for Whitney stratified spaces to obtain the structure of a stratifold, where the algebra consists of those functions commuting with appropriate representatives of the retractions). Here we only give a few simple examples. Consider $\mathbf{S} := \{(x, y) \in \mathbb{R}^2 \,|\, xy = 0\}$. We define $\mathbf{C}$ as the functions on $\mathbf{S}$ which are smooth away from $(0,0)$ and constant in some open neighbourhood of $(0,0)$. It is easy to show that $(\mathbf{S}, \mathbf{C})$ is a 1-dimensional stratifold with $\mathbf{S}^1 = \mathbf{S} - (0,0)$ and $\mathbf{S}^0 = (0,0)$.

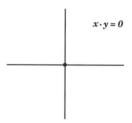

Example 4: In the same spirit we consider $\mathbf{S} := \{(x, y, z) \in \mathbb{R}^3 \,|\, x^2 + y^2 = z^2\}$. Again we define $\mathbf{C}$ as the functions on $\mathbf{S}$ which are smooth away from $(0,0,0)$ and constant in some open neighbourhood of $(0,0,0)$. This gives a 2-dimensional stratifold $(\mathbf{S}, \mathbf{C})$, where $\mathbf{S}^2 = \mathbf{S} - (0,0,0)$ and $\mathbf{S}^0 = (0,0,0)$.

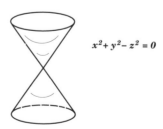

Example 5: Let $(\mathbf{S}, \mathbf{C})$ be a k-dimensional stratifold and $U \subset \mathbf{S}$ an open subset. Then $(U, \mathbf{C}(U))$ is a k-dimensional stratifold. We suggest that

3. Examples

the reader verify this to become acquainted with stratifolds.

Example 6: Let $(\mathbf{S}, \mathbf{C})$ and $(\mathbf{S}', \mathbf{C}')$ be stratifolds of dimension k and ℓ. Then we define a stratifold with underlying topological space $\mathbf{S} \times \mathbf{S}'$. To do this we use the local retractions (Proposition 2.1). We define $\mathbf{C}(\mathbf{S} \times \mathbf{S}')$ as those continuous functions $f : \mathbf{S} \times \mathbf{S}' \longrightarrow \mathbb{R}$ which are smooth on all products $\mathbf{S}^i \times (\mathbf{S}')^j$ and such that for each $(x, y) \in \mathbf{S}^i \times (\mathbf{S}')^j$ there are local retractions $r_x : U_x \longrightarrow \mathbf{S}^i \cap U_x$ and $r_y : U_y \longrightarrow (\mathbf{S}')^j \cap U_y$ for which $f|_{U_x \times U_y} = f(r_x \times r_y)$. In short, we define $\mathbf{C}(\mathbf{S} \times \mathbf{S}')$ as those continuous maps which commute with the product of appropriate local retractions onto $\mathbf{S}^i$ and $(\mathbf{S}')^j$. The detailed argument that $(\mathbf{S} \times \mathbf{S}', \mathbf{C}(\mathbf{S} \times \mathbf{S}'))$ is a $(k+\ell)$-dimensional stratifold is a bit lengthy and not relevant for further reading and for that reason we provide it in Appendix A. Both projections are morphisms.

In particular, if $(\mathbf{S}', \mathbf{C}')$ is a smooth m-dimensional manifold M, then we have the product stratifold $(\mathbf{S} \times M, \mathbf{C}(\mathbf{S} \times M))$.

Example 7: Combining example 6 with the method for constructing example 1, we construct the open cone over a compact stratifold $(\mathbf{S}, \mathbf{C})$. The underlying space is again $\overset{\circ}{C}\mathbf{S}$. We consider the algebra $\mathbf{C} \subset C^0(\overset{\circ}{C}\mathbf{S})$ consisting of all functions in $C^0(\overset{\circ}{C}\mathbf{S})$ which are constant on some open neighbourhood U of the top of the cone pt and whose restriction to $\mathbf{S} \times (0, 1)$ is in $\mathbf{C}(\mathbf{S} \times (0, 1))$. By arguments similar to those used for the cone over a compact manifold, one shows that $(\overset{\circ}{C}\mathbf{S}, \mathbf{C})$ is a $(k+1)$-dimensional stratifold.

Example 8: If $(\mathbf{S}, \mathbf{C})$ and $(\mathbf{S}', \mathbf{C}')$ are k-dimensional stratifolds, we define the **topological sum** whose underlying topological space is the disjoint union $\mathbf{S} \sqcup \mathbf{S}'$ (which is by definition $\mathbf{S} \times \{0\} \bigcup \mathbf{S}' \times \{1\}$) and whose algebra is given by those functions whose restriction to $\mathbf{S}$ is in $\mathbf{C}$ and to $\mathbf{S}'$ is in $\mathbf{C}'$. It is obvious that this is a k-dimensional stratifold.

Example 9: The following construction allows an inductive construction of stratifolds. We will not use it in this book (so the reader can skip it), but it provides a rich class of stratifolds. Let $(\mathbf{S}, \mathbf{C})$ be an n-dimensional stratifold and W a k-dimensional smooth manifold with boundary together with a collar $\mathbf{c} : \partial W \times [0, \epsilon) \to W$. We assume that $k > n$. Let $f : \partial W \to \mathbf{S}$ be a morphism, which we call the attaching map. We further assume that the attaching map f is proper, which in our context is equivalent to requiring that the preimages of compact sets are compact. Then we define a new space $\mathbf{S}'$ by gluing W to $\mathbf{S}$ via f:

$$\mathbf{S}' := W \cup_f \mathbf{S}.$$

On this space, we consider the algebra $\mathbf{C}'$ consisting of those functions $g : \mathbf{S}' \to \mathbb{R}$ whose restriction to $\mathbf{S}$ is in $\mathbf{C}$, whose restriction to the interior of W, $\overset{\circ}{W} := W - \partial W$ is smooth, and such that for some $\delta < \epsilon$ we have $g\mathbf{c}(x, t) = gf(x)$ for all $x \in \partial W$ and $t < \delta$. We leave it to the reader to check that $(\mathbf{S}', \mathbf{C}')$ is a k-dimensional stratifold. If $\mathbf{S}$ consists of a single point, we obtain a stratifold whose underlying space is $W/\partial W$, the space obtained by collapsing ∂W to a point. If W is compact and we apply this construction, then the result agrees with the stratifold from example 2 for $\overset{\circ}{W}$. Specializing further to $W := M \times [0, 1)$, where M is a closed manifold, we obtain the stratifold from example 1, the open cone over M.

Applying this construction inductively to a finite sequence of i-dimensional smooth manifolds W_i with compact boundaries equipped with collars and morphisms $f_i : \partial W_i \to \mathbf{S}^{i-1}$, where $\mathbf{S}^{i-1}$ is inductively constructed from $(W_0, f_0), \ldots, (W_{i-1}, f_{i-1})$, we obtain a rich class of stratifolds. Most stratifolds occurring in "nature" are of this type. This construction is very similar to the definition of CW-complexes. There we inductively attach cells (= closed balls), whereas here we attach arbitrary manifolds. Thus on the one hand it is more general, but on the other hand more special, since we require that the attaching maps are morphisms.

In this context it is sometimes useful to remember the data in this construction: the collars and the attaching maps. More precisely we pass from the collars to equivalence classes of collars called germs of collars, where two collars are equivalent if they agree on some small neighbourhood of the boundary. Stratifolds constructed inductively by attaching manifolds together using the data: germs of collars and attaching maps, are called **parametrized stratifolds** or *p*-**stratifolds**.

Example 10: It is not surprising that the same space can carry different structures as stratifolds. For example, the open cone over S^n is homeomorphic to the open disc B^{n+1}, which is on the one hand a smooth manifold and on the other hand by the construction of the cone above a stratifold with two non-empty strata, the point 0 and the rest. There are in addition very natural structures on B^{n+1} as a differential space, which don't give stratifolds. For example consider the algebra of smooth functions on B^{n+1} whose derivative at 0 is zero. The tangent space at a point $x \neq 0$ is $\mathbb{R}^{n+1}$, whereas the tangent space at 0 is zero. Thus we obtain a decomposition into two strata as in the case of the cone, namely 0 and the rest. But this is not a stratifold since the germ at 0 contains non locally constant functions.

From now on, we often omit the algebra C from the notation of a stratifold and write S instead of (S, C) (unless we want to make the dependence on C visible). This is in analogy to smooth manifolds where the single letter M is used instead of adding the maximal atlas or, equivalently, the algebra of smooth functions to the notation.

4. Properties of smooth maps

By analogy to maps from a smooth manifold to a smooth manifold, we call the morphisms f from a stratifold **S** to a smooth manifold **smooth maps**.

We now prove some elementary properties of smooth maps.

Proposition 2.2. *Let* **S** *be a stratifold and* $f_i : \mathbf{S} \to \mathbb{R}$ *be a family of smooth maps such that* $\operatorname{supp} f_j$ *is a locally finite family of subsets of* **S**. *Then* $\sum f_i$ *is a smooth map.*

Proof: The local finiteness implies that for each $x \in \mathbf{S}$, there is a neighbourhood U of x such that $\operatorname{supp} f_i \cap U = \varnothing$ for all but finitely many $i_1, \ldots, i_k$. Then it is clear that $\sum f_i|_U = f_{i_1}|_U + \cdots + f_{i_k}|_U$. Since $f_{i_1} + \cdots + f_{i_k}$ is smooth, we conclude from the fact that the algebra of smooth functions on **S** is locally detectable, that the map is smooth.
q.e.d.

We will now construct an important tool from differential topology, namely the existence of subordinate partitions of unity. This will make the role of the bump functions clear.

Recall that a **partition of unity** is a family of functions $\{\rho_i : \mathbf{S} \to \mathbb{R}_{\geq 0}\}$ such that their supports form a locally finite covering of **S** and $\sum \rho_i = 1$. It is called **subordinate** to some covering of **S**, if for each i the support $\operatorname{supp} \rho_i$ is contained in one of the covering sets.

Proposition 2.3. *Let* **S** *be a stratifold with an open covering. Then there is a subordinate partition of unity consisting of smooth functions (called a smooth partition of unity).*

Proof: The argument is similar to that for smooth manifolds [**B-J**, p. 66]. We choose a sequence of compact subspaces $A_i \subset \mathbf{S}$ such that $A_i \subset \mathring{A}_{i+1}$

and $\bigcup A_i = \mathbf{S}$. Such a sequence exists since $\mathbf{S}$ is locally compact and has a countable basis [**Sch**, p. 81)]. For each $x \in A_{i+1} - \mathring{A}_i$ we choose U from our covering such that $x \in U$ and take a smooth bump function $\rho_x^i : \mathbf{S} \to \mathbb{R}_{\geq 0}$ with $\operatorname{supp} \rho_x^i \subset (\mathring{A}_{i+2} - A_{i-1}) \cap U$. Since $A_{i+1} - \mathring{A}_i$ is compact, there is a finite number of points x_ν such that $(\rho_{x_\nu}^i)^{-1}(0, \infty)$ covers $A_{i+1} - \mathring{A}_i$. From Proposition 2.2 we know that $s := \sum_{i,\nu} \rho_{x_\nu}^i$ is a smooth, nowhere zero function and $\{\rho_{x_\nu}^i / s\}$ is the desired subordinate partition of unity.
q.e.d.

As a consequence, we note that $\mathbf{S}$ is a paracompact space.

To demonstrate the use of this result, we give the following standard applications.

Proposition 2.4. *Let $A \subset \mathbf{S}$ be a closed subset of a stratifold $\mathbf{S}$, U an open neighbourhood of A and $f : U \to \mathbb{R}$ a smooth function. Then there is a smooth function $g : \mathbf{S} \to \mathbb{R}$ such that $g|_A = f|_A$.*

Proof: The subsets U and $\mathbf{S} - A$ form an open covering of $\mathbf{S}$. Consider a subordinate smooth partition of unity $\{\rho_i : \mathbf{S} \to \mathbb{R}_{\geq 0}\}$. Then for $x \in U$ we define
$$g(x) := \sum_{\operatorname{supp} \rho_i \subset U} \rho_i(x) f(x),$$
and for $x \notin U$ we put $g(x) = 0$.
q.e.d.

Here is another useful consequence.

Proposition 2.5. *Let Y be a closed subspace of $\mathbf{S}$. Then $\mathbf{C}(Y)$ is equal to the restrictions of elements of $\mathbf{C}$ to Y.*

Proof: By definition $f : Y \to \mathbb{R}$ is in $\mathbf{C}(Y)$ if and only if for each $y \in Y$ there is a function $g_y \in \mathbf{C}$ and an open neighbourhood U_y of y in $\mathbf{S}$ such that $f|_{U_y \cap Y} = g|_{U_y \cap Y}$. Since Y is closed, the subsets U_y for $y \in Y$ and $\mathbf{S} - Y$ form an open covering of $\mathbf{S}$. Let $\{\rho_i : \mathbf{S} \to \mathbb{R}\}$ be a subordinate smooth partition of unity. Then for each i there is a $y(i)$ such that $\operatorname{supp} \rho_i \subset U_{y(i)}$ or $\operatorname{supp} \rho_i \subseteq \mathbf{S} - Y$. We consider the smooth function defined on Y
$$F := \sum_{\operatorname{supp} \rho_i \subset U_{y(i)}} \rho_i g_{y(i)}.$$

For $z \in Y$ we have
$$F(z) = \sum_{\text{supp } \rho_i \subset U_{y(i)}} \rho_i(z) g_{y(i)}(z) = \sum_i \rho_i(z) f(z) = f(z).$$
Here we have used that for $z \in Y$, if supp $rho_i \subset S \setminus Y$ then $\rho_i(z) = 0$, and that if $\rho_i(z) \neq 0$ then $g_{y(i)}(z) = f(z)$.
q.e.d.

5. Consequences of Sard's Theorem

One of the most useful fundamental results in differential topology is Sard's Theorem [**B-J**, p. 58], [**Hi**, p. 69], which implies that the regular values of a smooth map are dense (Brown's Theorem). As an immediate consequence of Sard's theorem for manifolds, we obtain a generalization of Brown's Theorem to stratifolds.

We recall that if $f : M \to N$ is a smooth map between smooth manifolds, then $x \in N$ is called a regular value of f if the differential df_y is surjective for each $y \in f^{-1}(x)$.

Definition: *Let $f : \mathbf{S} \to M$ be a smooth map from a stratifold to a smooth manifold. We say that $x \in M$ is a **regular value** of f, if for all $y \in f^{-1}(x)$ the differential df_y is surjective, or, equivalently, if x is a regular value of $f|_{\mathbf{S}^i}$ for all i.*

The equivalence of the two conditions comes from the fact that the tangent spaces of x in $\mathbf{S}^i$ and in $\mathbf{S}$ agree and also the differentials of f and $f|_{\mathbf{S}^i}$ at x are the same.

Let $f : M \to N$ be a smooth map between smooth manifolds. The image of a point $y \in M$ where the differential is not surjective is called a critical value. Sard's theorem says that the set of critical values has measure zero. This implies that its complement, the set of regular values, is dense (Brown's theorem). Since a finite union of sets of measure zero has measure zero, we deduce the following generalization of Brown's Theorem:

Proposition 2.6. *Let $g : \mathbf{S} \to M$ be a smooth map. The set of regular values of g is dense in M.*

Regular values x of smooth maps $f : M \to N$ have the useful property that $f^{-1}(x)$, the set of solutions, is a smooth manifold of dimension $\dim M - \dim N$. An analogous result holds for a smooth map $g : \mathbf{S} \to M$, where $\mathbf{S}$

is a stratifold of dimension n and M a smooth manifold without boundary of dimension m. Consider a regular value $x \in M$. By 2.5 we can identify $\mathbf{C}_{g^{-1}(x)}$ with the restriction of the smooth functions of $\mathbf{S}$ to $g^{-1}(x)$.

Proposition 2.7. *Let $\mathbf{S}$ be a k-dimensional stratifold, M an m-dimensional smooth manifold, $g : \mathbf{S} \to M$ be a smooth map and $x \in M$ a regular value. Then $(g^{-1}(x), \mathbf{C}(g^{-1}(x)))$ is a $(k - m)$-dimensional stratifold.*

Proof: We note that for each $y \in g^{-1}(x)$ the differential $dg_y : T_y\mathbf{S} \to T_xM$ as defined at the end of chapter 1 is surjective. This uses the property that $T_y\mathbf{S} = T_y\mathbf{S}^i$ if $y \in \mathbf{S}^i$. From this we conclude that $\dim T_y g^{-1}(x) \leq \dim T_y\mathbf{S} - m$. On the other hand, $T_y g^{-1}(x)$ contains $T_y((g|_{\mathbf{S}^i})^{-1}(x))$ and so the dimensions must be equal:
$$\dim T_y g^{-1}(x) = \dim T_y\mathbf{S} - m.$$
Thus $g^{-1}(x)^{i-m} = (g|_{\mathbf{S}^i})^{-1}(x)$, the stratification being induced from the stratification of $\mathbf{S}$.

The topological conditions of a stratifold are obvious. To show condition 1, we have to prove that
$$\mathbf{C}(g^{-1}(x))_y \to C^\infty(g^{-1}(x)^{i-m})_y$$
$$[f] \mapsto [f|_{g^{-1}(x)^{i-m}}]$$
is an isomorphism. We give an inverse by applying Proposition 2.1 to choose a local retraction $r : U \to V$ of $\mathbf{S}$ near y. The morphism gr is a local extension of $g|_V$ and $g|_U$ is another extension implying, by condition 1, that there exists a neighbourhood U' of y such that $gr|_{U'} = g|_{U'}$. Thus $r|_{g^{-1}(x) \cap U'} : g^{-1}(x) \cap U' \to g^{-1}(x)^{i-m}$ is a morphism. Now we obtain an inverse of $\mathbf{C}(g^{-1}(x))_y \to C^\infty(g^{-1}(x)^{i-m})_y$ by mapping $[f] \in C^\infty(g^{-1}(x)^{i-m})_y$ to $fr|_{g^{-1}(x) \cap U'}$. We have to show that $[fr|_{g^{-1}(x) \cap U'}]$ is in $\mathbf{C}(g^{-1}(x))_y$, i.e., is the restriction of an element of $\mathbf{C}_y$. But since $g^{-1}(x)^{i-m}$ is a smooth submanifold of $\mathbf{S}^i$, we can extend $[f]$ to a germ $[\hat{f}] \in (\mathbf{S}^i)_y$ and so $[\hat{f}r]$ is in $\mathbf{C}_y$ and $[\hat{f}r|_{g^{-1}(x)}] = [fr|_{g^{-1}(x) \cap U'}]$. Since $r|_{g^{-1}(x) \cap U'}$ is a local retraction, the map $[f] \mapsto [fr|_{g^{-1}(x) \cap U'}]$ is an inverse of $\mathbf{C}(g^{-1}(x))_y \to C^\infty(g^{-1}(x)^{i-m})_y$.

This establishes condition 1. The second condition is obvious and for condition 3 we note that bump functions are given by restriction of appropriate bump functions on $\mathbf{S}$.
q.e.d.

The next result will be very useful in proving properties of homology. It answers the following natural question. Let $\mathbf{S}$ be a connected k-dimensional stratifold and A and B non-empty disjoint closed subsets of **S**.

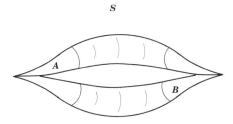

The question is whether there is a non-empty $(k-1)$-dimensional stratifold **S**$'$ with underlying topological space **S**$' \subset$ **S**$-(A \cup B)$ as in the following picture. If so, we say that **S**$'$ separates A and B in **S**.

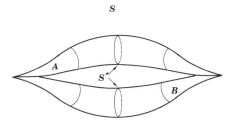

The positive answer uses several of the results presented so far. We first note that there is a smooth function $\rho :$ **S** $\to \mathbb{R}$ which maps A to 1 and B to -1. Namely, since **S** is paracompact, it is normal [**Sch**, p. 95] and thus there are disjoint open neighbourhoods U of A and V of B. Defining f as 1 on U and -1 on V, the existence of ρ follows from Proposition 2.4.

Now we apply Proposition 2.6 to see that the regular values of ρ are dense. Thus we can choose a regular value $t \in (-1, 1)$. Proposition 2.7 implies that $\rho^{-1}(t)$ is a non-empty $(k-1)$-dimensional stratifold which separates A and B. Thus we have proved a separation result:

Proposition 2.8. *Let **S** be a k-dimensional connected stratifold and A and B disjoint closed non-empty subsets of **S**. Then there is a non-empty $(k-1)$-dimensional stratifold **S**$'$ with **S**$' \subset$ **S**$-(A \cup B)$. That is, **S**$'$ separates A and B in **S**.*

6. Exercises

(1) Give a stratifold structure on the real plane $\mathbb{R}^2$ with 1-stratum equal to the x-axis and 2-stratum its complement. Verify that all the axioms of a stratifold hold.

(2) Let $(\mathbb{R}, \mathbf{C})$ be the real line with the algebra of smooth functions which are constant for $x \geq 0$. Is it a differential space? Is it a stratifold?

(3) Define $f : \mathbb{R} \to \mathbb{R}$ by $f(x) = 0$ for $x \leq 0$ and $f(x) = xe^{-\frac{1}{x^2}}$ for $x > 0$. Note that f is smooth. Since f is monotone for $0 \leq x$, it has an inverse $f^{-1} : [0, \infty) \to [0, \infty)$ which is continuous and, when restricted to $(0, \infty)$, it is smooth. Define a function $F : [0, \infty) \times [0, \infty) \to \mathbb{R}$ by $F(x, y) = f(f^{-1}(x) - y)$. F has the following properties:
1) F is well defined and continuous.
2) It is smooth when restricted to $(0, \infty) \times (0, \infty)$.
3) $F(x, 0) = x$.
4) $F(x, y) = 0$ for $x \leq f(y)$.
We extend F to be $F : \mathbb{R} \times [0, \infty) \to \mathbb{R}$ by setting $F(x, y) = -F(-x, y)$ for negative x.
a) Verify that F is well defined, continuous and smooth when restricted to $\mathbb{R} \times (0, \infty)$, $F(x, 0) = x$, and $F(x, y) = 0$ for $|x| \leq f(y)$.
b) Denote by $\mathbf{C}$ the set of functions $g : \mathbb{R} \times [0, \infty) \to \mathbb{R}$ which are continuous, smooth when restricted to $\mathbb{R} \times (0, \infty)$, smooth when restricted to $\mathbb{R} \times \{0\}$ (considered as a manifold) and locally commute with F, that is $g(x, y) = g(F(x, y), 0)$ for some neighborhood of the x-axis. Show that $(\mathbb{R} \times [0, \infty), \mathbf{C})$ is a differential space and a stratifold of dimension 2.

(4) Let $(\mathbb{R}^2, \mathbf{C})$ be the real plane with the algebra of smooth functions which are constant for $x \geq 0$. Is it a differential space? Is it a stratifold?

(5) Give a differential structure on the Hawaiian earring and on its spherical analog, having two strata.
The Hawaiian earring in $\mathbb{R}^2$ is the subspace $H = \bigcup S(\frac{1}{n}, (\frac{1}{n}, 0))$, where $S(r, (x, y))$ is the circle of radius r around the point (x, y). Note that all these circles have a common point which is $(0, 0)$. The spherical analog is the subspace of $\mathbb{R}^3$ which is the union of spheres instead of circles and is defined in a similar way.

(6) Let M be a manifold of dimension n with boundary and a collar. Show that it has a structure of a stratifold with two strata. (Hint: This is a special case of one of the examples below.)

(7) Let $(\mathbf{S}, \mathbf{C}), (\mathbf{S}', \mathbf{C}')$ be two stratifolds. Give a stratifold structure on the following topological spaces:
a) $\Sigma \mathbf{S}$, the suspension of $\mathbf{S}$ where $\Sigma \mathbf{S} = \mathbf{S} \times I / \sim$ where we identify $(a, 0) \sim (a', 0)$ and also $(a, 1) \sim (a', 1)$ for all $a, a' \in \mathbf{S}$.
b) More generally, on the join of $\mathbf{S}$ and $\mathbf{S}'$ which is denoted by

$\mathbf{S} * \mathbf{S}' := \mathbf{S} \times \mathbf{S}' \times I / \sim$ where we identify $(a, b, 0) \sim (a, b', 0)$ and also $(a, b, 1) \sim (a', b, 1)$ for all $a, a' \in \mathbf{S}$ and $b, b' \in \mathbf{S}'$ (the suspension is a special case since $\Sigma \mathbf{S} = \mathbf{S} * S^0$).

(8) a) Let $(\mathbf{S}, \mathbf{C})$ be a k-dimensional stratifold and let $f : \widetilde{\mathbf{S}} \to \mathbf{S}$ be a covering map. Show that there is a unique way to define a k-dimensional stratifold structure on $\widetilde{\mathbf{S}}$ such that f will be a local isomorphism.

b) Let $(\mathbf{S}, \mathbf{C})$ be a k-dimensional stratifold. Assume that a finite group acts on $\mathbf{S}$ via morphisms. Show that if the action is free the quotient space $\mathbf{S}/G$ has a unique structure of a k-dimensional stratifold such that the quotient map is a local isomorphism.

(9) Let $(\mathbf{S}, \mathbf{C})$ be a k-dimensional stratifold. Show that the inclusion map of each stratum $f : \mathbf{S}^i \hookrightarrow \mathbf{S}$ is a morphism and the differential df_x is an isomorphism for all $x \in \mathbf{S}^i$.

(10) Show that the composition of morphisms is again a morphism.

(11) Prove the statement from the second section that for a stratifold $(\mathbf{S}, \mathbf{C})$ a map f is in $\mathbf{C}$ if and only if $f|_{\mathbf{S}^i} \in \mathbf{C}(\mathbf{S}^i)$ for all i and it commutes with local retractions.

(12) Show that the map $C^\infty(\mathbf{S}^i)_x \to \mathbf{C}_x$ given by a local retraction is an inverse to the restriction map $\mathbf{C}_x \to C^\infty(\mathbf{S}^i)_x$.

(13) Let $(\mathbf{S}, \mathbf{C})$ be a k-dimensional stratifold and $U \subseteq \mathbf{S}$ an open subset. Show that the induced structure on U gives a stratifold structure on U.

(14) Show that the cone over a p-stratifold can be given a p-stratifold structure.

Chapter 3

Stratifolds with boundary: c-stratifolds

Stratifolds are generalizations of smooth manifolds without boundary, but we also want to be able to define stratifolds with boundary. To motivate the idea of this definition, we recall that a smooth manifold W with boundary has a collar, which is a diffeomorphism $\mathbf{c} : \partial W \times [0, \epsilon) \to V$, where V is an open neighbourhood of ∂W in W, and $\mathbf{c}|_{\partial W} = \mathrm{id}_{\partial W}$. Collars are useful for many constructions such as gluing manifolds with diffeomorphic boundaries together. This makes it plausible to add a collar to the definition of a manifold with boundary as additional structure. Actually it is enough to consider a germ of collars. We call a smooth manifold together with a germ of collars a c-**manifold**. Our stratifolds with boundary will be defined as stratifolds together with a germ of collars, and so we call them c-**stratifolds**.

Staying with smooth manifolds for a while, we observe that we can define manifolds which are equipped with a collar as follows. We consider a topological space W together with a closed subspace ∂W. We denote $W - \partial W$ by $\overset{\circ}{W}$ and call it the interior. We assume that $\overset{\circ}{W}$ and ∂W are smooth manifolds of dimension n and $n - 1$.

Definition: *Let $(W, \partial W)$ be a pair as above. A* **collar** *is a homeomorphism*

$$\mathbf{c} : U_\epsilon \to V,$$

where $\epsilon > 0$, $U_\epsilon := \partial W \times [0, \epsilon)$, and V is an open neighbourhood of ∂W in W such that $c|_{\partial W \times \{0\}}$ is the identity map to ∂W and $\mathbf{c}|_{U-(\partial W \times \{0\})}$ is a *diffeomorphism onto $V - \partial W$.*

The condition requiring that $\mathbf{c}(U_\epsilon)$ is open avoids the following situation:

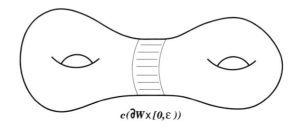

$c(\partial W \times [0,\varepsilon))$

Namely, it guarantees that the image of $\mathbf{c}$ is an "end" of W.

What is the relation to smooth manifolds equipped with a collar? If W is a smooth manifold and $\mathbf{c}$ a collar, then we obviously obtain all the ingredients of the definition above by considering W as a topological space. In turn, if $(W, \partial W, \mathbf{c})$ is given as in the definition above, we can in an obvious way extend the smooth structure of $\overset{\circ}{W}$ to a smooth manifold W with boundary. The smooth structure on W is characterized by requiring that $\mathbf{c}$ is not only a homeomorphism but a diffeomorphism. The advantage of the definition above is that it can be given using only the language of manifolds without boundary. Thus it can be generalized to stratifolds.

Let $(\mathbf{T}, \partial \mathbf{T})$ be a pair of topological spaces. We denote $\mathbf{T} - \partial \mathbf{T}$ by $\overset{\circ}{\mathbf{T}}$ and call it the interior. We assume that $\overset{\circ}{\mathbf{T}}$ and $\partial \mathbf{T}$ are stratifolds of dimension n and $n-1$ and that $\partial \mathbf{T}$ is a closed subspace.

Definition: *Let $(\mathbf{T}, \partial \mathbf{T})$ be a pair as above. A **collar** is a homeomorphism*

$$\mathbf{c}: U_\epsilon \to V,$$

where $\epsilon > 0$, $U_\epsilon := \partial \mathbf{T} \times [0, \epsilon)$, and V is an open neighbourhood of $\partial \mathbf{T}$ in $\mathbf{T}$ such that $\mathbf{c}|_{\partial \mathbf{T} \times \{0\}}$ is the identity map to $\partial \mathbf{T}$ and $\mathbf{c}|_{U_\epsilon - (\partial \mathbf{T} \times \{0\})}$ is an isomorphism of stratifolds onto $V - \partial \mathbf{T}$.

Perhaps this definition needs some explanation. By examples 5 and 6 in §1 the open subset $U_\epsilon - (\partial \mathbf{T} \times \{0\})$ can be considered as a stratifold. Similarly, $V - \partial \mathbf{T}$ is an open subset of $\mathbf{T}$ and thus, by example 5 in §2, it can be considered as a stratifold.

We are only interested in a germ of collars, which is an equivalence class of collars where two collars $\mathbf{c}: U_\epsilon \to V$ and $\mathbf{c}': U'_{\epsilon'} \to V'$ are called equivalent if there is a $\delta < \min\{\epsilon, \epsilon'\}$, such that $\mathbf{c}|_{U_\delta} = \mathbf{c}'|_{U_\delta}$. As usual when we consider equivalence classes, we denote the germ represented by a collar $\mathbf{c}$ by $[\mathbf{c}]$.

Now we define:

Definition: *An n-dimensional c-**stratifold** $\mathbf{T}$ (a collared stratifold) is a pair of topological spaces $(\mathbf{T}, \partial \mathbf{T})$ together with a germ of collars $[\mathbf{c}]$ where $\overset{\circ}{\mathbf{T}} = \mathbf{T} - \partial \mathbf{T}$ is an n-dimensional stratifold and $\partial \mathbf{T}$ is an $(n-1)$-dimensional stratifold, which is a closed subspace of $\mathbf{T}$. We call $\partial \mathbf{T}$ the **boundary** of $\mathbf{T}$.*

*A **smooth map** from $\mathbf{T}$ to a smooth manifold M is a continuous function f whose restriction to $\overset{\circ}{\mathbf{T}}$ and to $\partial \mathbf{T}$ is smooth and which commutes with an appropriate representative of the germ of collars, i.e., there is a $\delta > 0$ such that $f\mathbf{c}(x,t) = f(x)$ for all $x \in \partial \mathbf{T}$ and $t < \delta$.*

We often call $\mathbf{T}$ the underlying space of the c-stratifold.

As for manifolds, we allow $\partial \mathbf{T}$ to be empty. Then, of course, a c-stratifold is nothing but a stratifold without boundary (or better with an empty boundary). In this way stratifolds are incorporated into the world of c-stratifolds as those c-stratifolds $\mathbf{T}$ with $\partial \mathbf{T} = \emptyset$.

The simplest examples of c-stratifolds are given by c-manifolds W. Here we define $\mathbf{T} = W$ and $\partial \mathbf{T} = \partial W$ and attach to $\overset{\circ}{\mathbf{T}}$ and $\partial \mathbf{T}$ the stratifold and collar structures given by the smooth manifolds. Another important class of examples is given by the product of a stratifold $\mathbf{S}$ with a c-manifold W. By this we mean the c-stratifold whose underlying topological space is $\mathbf{S} \times W$, whose interior is $\mathbf{S} \times \overset{\circ}{W}$ and whose boundary is $\mathbf{S} \times \partial W$, and whose germ of collars is represented by $\mathrm{id}_\mathbf{S} \times \mathbf{c}$, where $[\mathbf{c}]$ is the germ of collars of W. We abbreviate this c-stratifold by $\mathbf{S} \times W$. In particular, we obtain the half open cylinder $\mathbf{S} \times [0,1)$ or the **cylinder** $\mathbf{S} \times [0,1]$. A third simple class of c-stratifolds is obtained by the product of a c-stratifold $\mathbf{T}$ with a smooth manifold M. The underlying topological space of this stratifold is given by $\mathbf{T} \times M$ with interior $\overset{\circ}{\mathbf{T}} \times M$ and boundary $\partial \mathbf{T} \times M$ and germ of collars $[\mathbf{c} \times \mathrm{id}_M]$, where $[\mathbf{c}]$ is the germ of collars of $\mathbf{T}$.

The next example is the **(closed) cone** $C(\mathbf{S})$ over a stratifold $\mathbf{S}$. The underlying topological space is the (closed) cone $\mathbf{T} := \mathbf{S} \times [0,1]/_{\mathbf{S}\times\{0\}}$ whose interior is $\mathbf{S} \times [0,1)/_{\mathbf{S}\times\{0\}}$ and whose boundary is $\mathbf{S} \times \{1\}$. The collar is given by the map $\mathbf{S} \times [0, 1/2) \to C(\mathbf{S})$ mapping (x,t) to $(x, 1-t)$.

In contrast to manifolds with boundary, where the boundary can be recognized from the underlying topological space, this is not the case with c-stratifolds. For example we can consider a c-manifold W as a stratifold without boundary with algebra $\mathbf{C}$ given by the functions which are smooth on the boundary and interior and commute with the retraction given by a representative of the germ of collars. Here the strata are the boundary and the interior of W. On the other hand it is—as mentioned above—a c-stratifold with boundary ∂W. In both cases the smooth functions agree.

The following construction of cutting along a codimension-1 stratifold will be useful later on. Suppose in the situation of Proposition 2.7, where $g : \mathbf{S} \to \mathbb{R}$ is a smooth map to the reals with regular value t, that there is an open neighbourhood U of $g^{-1}(t)$ and an isomorphism from $g^{-1}(t) \times (t-\epsilon, t+\epsilon)$ to U for some $\epsilon > 0$, whose restriction to $g^{-1}(t) \times \{0\}$ is the identity map to $g^{-1}(t)$. Such an isomorphism is often called a **bicollar**. Then we consider the spaces $\mathbf{T}_+ := g^{-1}[t, \infty)$ and $\mathbf{T}_- := g^{-1}(-\infty, t]$. We define their boundary as $\partial \mathbf{T}_+ := g^{-1}(t)$ and $\partial \mathbf{T}_- := g^{-1}(t)$. Since $\overset{\circ}{\mathbf{T}}_+$ and $\overset{\circ}{\mathbf{T}}_-$ are open subsets of $\mathbf{S}$ they are stratifolds. The restriction of the isomorphism to $g^{-1}(t) \times [t, t+\epsilon)$ is a collar of $\mathbf{T}_+$ and the restriction of the isomorphism to $g^{-1}(t) \times (t-\epsilon, t]$ is a collar of $\mathbf{T}_-$. Thus we obtain two c-stratifolds $\mathbf{T}_+$ and $\mathbf{T}_-$. We say that $\mathbf{T}_+$ and $\mathbf{T}_-$ are obtained from $\mathbf{S}$ by **cutting along a codimension-1 stratifold**, namely along $g^{-1}(t)$.

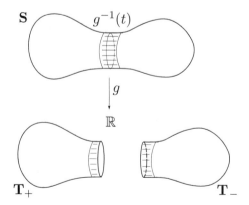

Now we describe the reverse process and introduce **gluing of stratifolds along a common boundary**. Let $\mathbf{T}$ and $\mathbf{T}'$ be c-stratifolds with the same boundary, $\partial \mathbf{T} = \partial \mathbf{T}'$. By passing to the minimum of ϵ and ϵ' we can assume that the domains of the collars are equal: $\mathbf{c} : \partial \mathbf{T} \times [0, \epsilon) \to V \subset \mathbf{T}$ and $\mathbf{c}' : \partial \mathbf{T}' \times [0, \epsilon) \to V' \subset \mathbf{T}'$. Then we consider the topological space $\mathbf{T} \cup_{\partial \mathbf{T} = \partial \mathbf{T}'} \mathbf{T}'$ obtained from the disjoint union of $\mathbf{T}$ and $\mathbf{T}'$ by identifying the boundaries. We have a bicollar (in the world of topological spaces), a homeomorphism $\varphi : \partial \mathbf{T} \times (-\epsilon, \epsilon) \to V \cup V'$ by mapping $(x, t) \in \partial \mathbf{T} \times (-\epsilon, 0]$ to $\mathbf{c}(x, -t)$ and $(x, t) \in \partial \mathbf{T} \times [0, \epsilon)$ to $\mathbf{c}'(x, t)$.

With respect to this underlying topological space, we define the algebra $\mathbf{C}(\mathbf{T} \cup_{\partial \mathbf{T} = \partial \mathbf{T}'} \mathbf{T}')$ to consist of those continuous maps $f : \mathbf{T} \cup_{\partial \mathbf{T} = \partial \mathbf{T}'} \mathbf{T}' \to \mathbb{R}$, such that the restrictions to $\overset{\circ}{\mathbf{T}}$ and $\overset{\circ}{\mathbf{T}}'$ are in $\mathbf{C}(\overset{\circ}{\mathbf{T}})$ and $\mathbf{C}(\overset{\circ}{\mathbf{T}}')$ respectively, and where the composition $f\varphi : \partial \mathbf{T} \times (-\epsilon, \epsilon) \to \mathbb{R}$ is in $\mathbf{C}(\partial \mathbf{T} \times (-\epsilon, \epsilon))$. It is easy to see that $\mathbf{C}(\mathbf{T} \cup_{\partial \mathbf{T} = \partial \mathbf{T}'} \mathbf{T}')$ is a locally detectable algebra. Since the condition in the definition of differential spaces is obviously fulfilled, we have a differential space. Clearly, $\mathbf{T} \cup_{\partial \mathbf{T} = \partial \mathbf{T}'} \mathbf{T}'$ is a locally compact Hausdorff space with countable basis. The conditions (1) – (3) in the definition of a stratifold are local conditions. Since they hold for $\overset{\circ}{\mathbf{T}}$, $\overset{\circ}{\mathbf{T}}'$, and $\partial \mathbf{T} \times (-\epsilon, \epsilon)$ and φ is an isomorphism, they hold for $\mathbf{T} \cup_{\partial \mathbf{T} = \partial \mathbf{T}'} \mathbf{T}'$. Thus $(\mathbf{T} \cup_{\partial \mathbf{T} = \partial \mathbf{T}'} \mathbf{T}', \mathbf{C}(\mathbf{T} \cup_{\partial \mathbf{T} = \partial \mathbf{T}'} \mathbf{T}'))$ is a stratifold.

One can generalize the context of the above construction by assuming only the existence of an isomorphism g: $\partial \mathbf{T} \to \partial \mathbf{T}'$ rather than $\partial \mathbf{T} = \partial \mathbf{T}'$. Then we glue the spaces via g to obtain a space $\mathbf{T} \cup_g \mathbf{T}'$. If in the definition of the algebra $\mathbf{C}(\mathbf{T} \cup_{\partial \mathbf{T} = \partial \mathbf{T}'} \mathbf{T}')$ we replace the homeomorphism φ by $\varphi_g : \partial \mathbf{T} \times (-\epsilon, \epsilon) \to V \cup V'$ mapping $(x, t) \in \partial \mathbf{T} \times (-\epsilon, 0]$ to $\mathbf{c}(x, -t)$ and $(x, t) \in \partial \mathbf{T} \times [0, \epsilon)$ to $\mathbf{c}'(g(x), t)$, then we obtain a locally detectable algebra $\mathbf{C}(\mathbf{T} \cup_g \mathbf{T}')$. The same arguments as above used for $g = $ id imply that $(\mathbf{T} \cup_g \mathbf{T}', \mathbf{C}(\mathbf{T} \cup_g \mathbf{T}'))$ is a stratifold. We summarize this as:

Proposition 3.1. *Let $\mathbf{T}$ and $\mathbf{T}'$ be k-dimensional c-stratifolds and let $g : \partial \mathbf{T} \to \partial \mathbf{T}'$ be an isomorphism. Then*

$$(\mathbf{T} \cup_g \mathbf{T}', \mathbf{C}(\mathbf{T} \cup_g \mathbf{T}'))$$

is a k-dimensional stratifold.

Of course, if g is an isomorphism between some components of the boundary of $\mathbf{T}$ and some components of the boundary of $\mathbf{T}'$, we can glue as above via g to obtain a c-stratifold, whose boundary is the union of the complements of these boundary components (see Appendix B, §2).

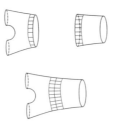

Finally we note that if $f : \mathbf{T} \to \mathbb{R}$ is a smooth function and s is a regular value of $f|_{\overset{\circ}{\mathbf{T}}}$ and $f|_{\partial \mathbf{T}}$, then $f^{-1}(s)$ is a c-stratifold with collar given by restriction.

1. Exercises

(1) Let $(\mathbf{T}, \partial \mathbf{T})$ be a compact c-stratifold of dimension n with an empty $n-1$ stratum such that the $(n-1)$-stratum of $\partial \mathbf{T}$ in non-empty. Show that $\partial \mathbf{T}$ is not a retract of $\mathbf{T}$, that is, there is no morphism $r : \mathbf{T} \to \partial \mathbf{T}$ which is the identity on $\partial \mathbf{T}$ and commutes with the collar. (Hint: Look at the preimage of a regular value in the top stratum.)

(2) Let $(\mathbf{S}, \partial \mathbf{S})$ and $(\mathbf{T}, \partial \mathbf{T})$ be two c-stratifolds. Construct two c-stratifolds having $\partial \mathbf{S} \times \partial \mathbf{T}$ as a boundary. What is the space obtained by gluing both stratifolds along the common boundary in the case of (D^{n+1}, S^n) and (D^{m+1}, S^m)?

(3) a) Let M be a smooth manifold and let $M_s = (M, \mathbf{C})$ be some stratifold structure on the topological space M. Show that $\dim M_s \leq \dim M$. Is the identity map $\mathrm{id} : M_s \to M$ in this case always a morphism?
b) Is there a c-stratifold structure on $(M \times I, \partial)$ with $\partial = M \sqcup M_s$ where M has the manifold structure and M_s is M with some stratifold structure?

Chapter 4

$\mathbb{Z}/2$-homology

Prerequisites: We use the classification of 1-dimensional compact manifolds (e.g. [**Mi 2**, Appendix]).

1. Motivation of homology

We begin by motivating the concept of a homology theory. In this book we will construct several homology theories which are all in the same spirit in the sense that they all attempt to measure the complexity of a space X by analyzing the "holes" in X. Initially, one thinks of a hole in a space as follows: let Y be a topological space and L a non-empty subspace. Then we say that $X := Y - L$ has the hole L. We call such a hole an **extrinsic hole** since we need to know the bigger space Y to say that X has a hole. We also want to say what it means that X has a hole without knowing Y. Such a hole we would call an **intrinsic hole**. The idea is rather simple: we try to detect holes by fishing for them with a net, which is some other space S. We throw (= map) the net into X and try to shrink the net to a point. If this is not possible, we have "caught" a hole. For example, if we consider $X = \mathbb{R}^n - 0$, then we would say that X is obtained from $\mathbb{R}^n$ by introducing the hole 0. We can detect the hole by mapping the "net" S^{n-1} to X via the inclusion. Since we cannot shrink S^{n-1} in X continuously to a point, we have "caught" the hole without using that X sits in $\mathbb{R}^n$.

This is a very flexible concept since we are free in choosing the shape of our net. In this chapter our nets will be certain compact stratifolds. Later we will consider other classes of stratifolds. Further flexibility will come from the fact that we can use stratifolds of different dimensions for detecting "holes of these dimensions". Finally, we are free in making precise what we mean by shrinking a net to a point. Here we will use a very rough criterion: we say that a net given by a map from a stratifold **S** to X can be "shrunk" to a point if there is a compact c-stratifold **T** with $\partial \mathbf{T} = \mathbf{S}$ and one can extend the map from **S** to **T**. In other words, instead of shrinking the net, we "fill" it with a compact stratifold **T**.

To explain this idea further, we start again from the situation where the space X is obtained from a space Y by deleting a set L. Depending on the choice of L, this may be a very strange space. Since we are more interested in nice spaces, let us assume that L is the interior of a compact c-stratifold $\mathbf{T} \subset Y$, i.e., $L = \overset{\circ}{\mathbf{T}}$. Then we can consider the inclusion of $\partial \mathbf{T}$ into $X = Y - L$ as our net. We say that this inclusion detects the hole obtained by deleting $\overset{\circ}{\mathbf{T}}$ if we cannot extend the inclusion from $\partial \mathbf{T}$ to X to a map from **T** into X.

We now weaken our knowledge of X by assuming that it is obtained from Y by deleting the interior of some compact c-stratifold, but we do not know which one. We only know the boundary **S** of the deleted c-stratifold. Then the only way to test if we have a hole with boundary—the boundary of the deleted stratifold—is to consider **all** compact c-stratifolds **T** having the same boundary **S** and to try to extend the inclusion of the boundary to a continuous map from **T** to X. If this is impossible for all **T**, then we say that X has a hole.

We have found a formulation of "hole" which makes sense for an arbitrary space X. The space X has a hole with the boundary shape of a compact stratifold **S** without boundary if there is an embedding of **S** into X which cannot be extended to any compact c-stratifold with boundary **S**. Furthermore, instead of fishing for holes with only embeddings, we consider arbitrary continuous maps from compact stratifolds **S** to X. We say that such a map catches a hole if we cannot extend it to a continuous map of any compact c-stratifold **T** with boundary **S**. Finally, we collect all the continuous maps from all compact stratifolds **S** of a fixed dimension m to X modulo those extending to a compact c-stratifold with boundary **S**, into a set. We find an obvious group structure on this set to obtain our first homology group; denoted $SH_m(X; \mathbb{Z}/2)$.

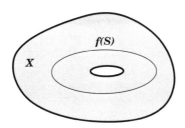

The idea for introducing homology this way is essentially contained in Poincaré's original paper from 1895 [**Po**]. Instead of using the concept of stratifolds, he uses objects called "variétés". The definition of these objects is not very clear in this paper, which leads to serious difficulties. As a consequence, he suggested another combinatorial approach which turned out to be successful, and is the basis of the standard approach to homology. The original idea of Poincaré was taken up by Thom [**Th 1**] around 1950 and later on by Conner and Floyd [**C-F**] who introduced a homology theory in the spirit of Poincaré's original approach using smooth manifolds. The construction of this homology theory is very easy but computations are much harder than for ordinary homology. In this book, we use these ideas with some technical modification to realize Poincaré's original idea in a textbook.

2. $\mathbb{Z}/2$-oriented stratifolds

We begin with the construction of our first homology theory by following the motivation above. The elements of our homology groups for a topological space X will be equivalence classes of certain pairs $(\mathbf{S}, g)$ consisting of an m-dimensional stratifold $\mathbf{S}$ together with a continuous map $g : \mathbf{S} \to X$. The equivalence relation is called bordism.

Before we define bordism, we must introduce the concept of an isomorphism between pairs $(\mathbf{S}, g)$, and $(\mathbf{S}', g')$.

Definition: *Let X be a topological space and $g : \mathbf{S} \to X$ and $g' : \mathbf{S}' \to X$ be continuous maps, where $\mathbf{S}$ and $\mathbf{S}'$ are m-dimensional stratifolds. An* **isomorphism** *from $(\mathbf{S}, g)$ to $(\mathbf{S}', g')$ is an isomorphism of stratifolds $f : \mathbf{S} \to \mathbf{S}'$ such that*
$$g = g'f.$$
If such an isomorphism exists, we call $(\mathbf{S}, g)$ and $(\mathbf{S}', g')$ isomorphic.

For a space X, the collection of pairs $(\mathbf{S}, g)$, where $\mathbf{S}$ is an m-dimensional stratifold and $g : \mathbf{S} \to X$ a continuous map, does not form a set. To see this, start with a fixed pair $(\mathbf{S}, g)$ and consider the pairs $(\mathbf{S} \times \{i\}, g)$, where

i is an arbitrary index. For example, we could take i to be any set. Thus, there are at least as many pairs as sets and the class of all sets is not a set. But we have

Proposition 4.1. *The isomorphism classes of pairs* $(\mathbf{S}, g)$ *form a set.*

The proof of this proposition does not help with the understanding of homology. Thus we have postponed it to the end of Appendix A (as we have done with other proofs which are more technical and whose understanding is not needed for reading the rest of the book).

We now introduce the operation which leads to homology groups. Given two pairs $(\mathbf{S}_1, g_1)$ and $(\mathbf{S}_2, g_2)$, their sum is

$$(\mathbf{S}_1, g_1) + (\mathbf{S}_2, g_2) := (\mathbf{S}_1 \sqcup \mathbf{S}_2, g_1 \sqcup g_2),$$

where $g_1 \sqcup g_2 : \mathbf{S}_1 \sqcup \mathbf{S}_2 \to X$ is the disjoint sum of the maps g_1 and g_2. If $\mathbf{T}$ is a c-stratifold and $f : \mathbf{T} \to X$ a map, we abbreviate

$$\partial(\mathbf{T}, f) := (\partial \mathbf{T}, f|_{\partial \mathbf{T}}).$$

We will now characterize those stratifolds from which we will construct our homology groups. There are two conditions we impose: $\mathbb{Z}/2$-orientability and regularity.

Definition: *We call an n-dimensional c-stratifold $\mathbf{T}$ with boundary $\mathbf{S} = \partial \mathbf{T}$ (we allow the possibility that $\partial \mathbf{T}$ is empty) $\mathbb{Z}/2$-oriented if $(\overset{\circ}{\mathbf{T}})^{n-1} = \varnothing$, i.e., if the stratum of codimension 1 is empty.*

We note that if $(\overset{\circ}{\mathbf{T}})^{n-1} = \varnothing$, then $\mathbf{S}^{n-2} = \varnothing$. The reason is that via $\mathbf{c}$ we have an embedding of $U = \mathbf{S}^{n-2} \times (0, \epsilon)$ into $(\overset{\circ}{\mathbf{T}})^{n-1}$ and so $U = \varnothing$ if $(\overset{\circ}{\mathbf{T}})^{n-1} = \varnothing$. But if $U = \varnothing$ then also $\mathbf{S}^{n-2} = \varnothing$. Thus the boundary of a $\mathbb{Z}/2$-oriented stratifold is itself $\mathbb{Z}/2$-oriented.

Remark: It is not clear at this moment what the notion "$\mathbb{Z}/2$-oriented" has to do with our intuitive imagination of orientation (knowing what is "left" and "right"). For a connected closed smooth manifold, we know what "oriented" means [**B-J**]. If M is a closed manifold, this concept can be translated to a homological condition using integral homology. It is equivalent to the existence of the so-called fundamental class. Our definition of $\mathbb{Z}/2$-oriented stratifolds guarantees that a closed smooth manifold always has a $\mathbb{Z}/2$-fundamental class as we shall explain later.

3. Regular stratifolds

We distinguish another class of stratifolds by imposing a further local condition.

Definition: *A stratifold $\mathbf{S}$ is called a **regular stratifold** if for each $x \in \mathbf{S}^i$ there is an open neighborhood U of x in $\mathbf{S}$, a stratifold $\mathbf{F}$ with $\mathbf{F}^0$ a single point* pt, *an open subset V of $\mathbf{S}^i$, and an isomorphism*
$$\varphi : V \times \mathbf{F} \to U,$$
whose restriction to $V \times$ pt is the identity.

*A c-stratifold $\mathbf{T}$ is called a **regular c-stratifold** if $\overset{\circ}{\mathbf{T}}$ and $\partial \mathbf{T}$ are regular stratifolds.*

To obtain a feeling for this condition, we look at some examples. We note that a smooth manifold is a regular stratifold. If $\mathbf{S}$ is a regular stratifold and M a smooth manifold, then $\mathbf{S} \times M$ is a regular stratifold. Namely for $(x, y) \in \mathbf{S} \times M$, we consider an isomorphism $\varphi : V \times \mathbf{F} \to U$ near x for $\mathbf{S}$ as above and then
$$\varphi \times \mathrm{id} : (V \times \mathbf{F}) \times M \to U \times M$$
is an isomorphism near (x, y). Thus $\mathbf{S} \times M$ is a regular stratifold. A similar consideration shows that the product $\mathbf{S} \times \mathbf{S}'$ of two regular stratifolds $\mathbf{S}$ and $\mathbf{S}'$ is regular.

Another example of a regular stratifold is the open cone over a compact smooth manifold M. More generally, if $\mathbf{S}$ is a regular stratifold, then the open cone $\overset{\circ}{C}\mathbf{S}$ is a regular stratifold: by the considerations above the open subset $\mathbf{S} \times (0, 1)$ is a regular stratifold and it remains to check the condition for the 0-stratum, but this is clear as we can take $U = \mathbf{F} = \overset{\circ}{C}\mathbf{S}$ and $V =$ pt.

It is obvious that the topological sum of two regular stratifolds is regular.

Thus the constructions of stratifolds using regular stratifolds from the examples in chapter 2 lead to regular stratifolds.

It is also obvious that gluing regular stratifolds together as explained in Proposition 3.1 leads to a regular stratifold. The reason is that points in the gluing look locally like points in either $\overset{\circ}{\mathbf{T}}$, $\overset{\circ}{\mathbf{T}}'$ or $\partial T \times (-\epsilon, \epsilon)$. Since these stratifolds are regular and regularity is a local condition, the statement follows.

Finally, if $\mathbf{S}$ is a regular stratifold and $f : \mathbf{S} \to \mathbb{R}$ is a smooth map with regular value t, then $f^{-1}(t)$ is a regular stratifold. To see this, it is enough to consider the local situation near x in $\mathbf{S}^i$ and use a local isomorphism φ to reduce to the case, where the stratifold is $V \times \mathbf{F}$ for some i-dimensional manifold V and $\mathbf{F}^0$ is a point pt. Now we consider the maps f and $(f|_{V \times \text{pt}}) p_1$, where p_1 is the projection to V, and note that they agree on $V \times \text{pt}$, which is the i-stratum of $V \times \mathbf{F}$. By condition 1b of a stratifold there is some open neighbourhood W of pt in $\mathbf{F}$ such that the maps agree on $V \times W$. Thus $f|_{V \times W} = (f|_{V \times \text{pt}}) p$, where p is the projection from $V \times W$ to V. Since t is a regular value of $f|_{V \times \text{pt}}$, we see that $f^{-1}(t) \cap (V \times W) = f|_{V \times \text{pt}}^{-1}(t) \times W$ showing that the conditions of a regular stratifold are fulfilled. Since we will apply this result in the next chapter, we summarize this as:

Proposition 4.2. *Let $\mathbf{S}$ be a regular stratifold, $f : \mathbf{S} \to \mathbb{R}$ a smooth function and t a regular value. Then $f^{-1}(t)$ is a regular stratifold.*

The main reason for introducing regular stratifolds in our context is the following result. A regular point of a smooth map $f : \mathbf{S} \to \mathbb{R}$ is a point x in $\mathbf{S}$ such that the differential at x is non-zero.

Proposition 4.3. *Let $\mathbf{S}$ be a regular stratifold. Then the regular points of a smooth map $f : \mathbf{S} \to \mathbb{R}$ form an open subset of $\mathbf{S}$. If in addition $\mathbf{S}$ is compact, the regular values form an open set.*

Proof: To see the first statement consider a regular point $x \in \mathbf{S}^i$. Since $\mathbf{S}^i$ is a smooth manifold and the regular points of a smooth map on a smooth manifold are open (use the continuity of the differential to see this), there is an open neighbourhood U of x in $\mathbf{S}^i$ consisting of regular points. Since $\mathbf{S}$ is regular, there is an open neighbourhood U_x of x in $\mathbf{S}$ isomorphic to $V \times \mathbf{F}$, where $V \subset U$ is an open neighbourhood of x in $\mathbf{S}^i$, such that f corresponds on $V \times \mathbf{F}$ to a map which commutes with the projection from $V \times \mathbf{F}$ to V (this uses the fact that a smooth map has locally a unique germ of extensions to an open neighbourhood). But for a map which commutes with this projection, a point $(x, y) \in V \times \mathbf{F}$ is a regular point if and only if x is a regular point of $f|_V$. Since V is contained in U and U consists of regular points, U_x also consists of regular points, which finishes the proof the first statement.

If the regular points are an open set then the singular points, which are the complement, are a closed set. If **S** is compact, the singular points are compact, and so the image under f is closed implying that the regular values are open.
q.e.d.

4. $\mathbb{Z}/2$-homology

We call a c-stratifold **T** **compact** if the underlying space **T** is compact. Since $\partial \mathbf{T}$ is a closed subset of **T**, the boundary of a compact regular stratifold is compact.

Definition: *Two pairs $(\mathbf{S}_0, g_0)$ and $(\mathbf{S}_1, g_1)$, where $\mathbf{S}_i$ are compact, m-dimensional $\mathbb{Z}/2$-oriented, regular stratifolds and $g_i : \mathbf{S}_i \to X$ are continuous maps, are called* **bordant** *if there is a compact $(m+1)$-dimensional $\mathbb{Z}/2$-oriented regular c-stratifold **T**, and a continuous map $g : \mathbf{T} \to X$ such that $(\partial \mathbf{T}, g) = (\mathbf{S}_0, g_0) + (\mathbf{S}_1, g_1)$. The pair $(\mathbf{T}, g)$ is called a* **bordism** *between $(\mathbf{S}_0, g_0)$ and $(\mathbf{S}_1, g_1)$.*

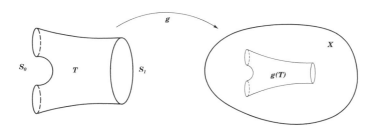

We will later see why we imposed the condition that the stratifolds are $\mathbb{Z}/2$-oriented and regular (the latter condition can be replaced by other conditions as long as an appropriate version of Proposition 4.3 holds). If we did not require that the c-stratifolds be compact, we would obtain a single bordism class, since otherwise we could use $(\mathbf{S} \times [0, \infty), gp)$ as a bordism between $(\mathbf{S}, g)$ and the empty stratifold (where p is the projection from $\mathbf{S} \times [0, \infty)$ to $\mathbf{S}$).

Proposition 4.4. *Let X be a topological space. Bordism defines an equivalence relation on the set of isomorphism classes of compact, m-dimensional, $\mathbb{Z}/2$-oriented, regular stratifolds with a map to X. Moreover, the topological sum*

$$(\mathbf{S}_0, g_0) + (\mathbf{S}_1, g_1) := (\mathbf{S}_0 \sqcup \mathbf{S}_1, g_0 \sqcup g_1)$$

induces the structure of an abelian group on the set of such equivalence classes. This group is denoted by $SH_m(X; \mathbb{Z}/2)$, the m-th stratifold homology group with $\mathbb{Z}/2$-coefficients or for short m-th $\mathbb{Z}/2$-homology. As usual, we denote the equivalence class represented by $(\mathbf{S}, g)$ by $[\mathbf{S}, g]$.

We will show in chapter 20 that for many spaces stratifold homology agrees with the most common and most important homology groups: **singular homology groups**.

Proof: $(\mathbf{S}, g)$ is bordant to $(\mathbf{S}, g)$ via the bordism $(\mathbf{S} \times [0,1], h)$, where $h(x,t) = g(x)$. We call this bordism the cylinder over $(\mathbf{S}, g)$. We observe that if $\mathbf{S}$ is $\mathbb{Z}/2$-oriented and regular, then $\mathbf{S} \times [0,1]$ is $\mathbb{Z}/2$-oriented and regular. Thus the relation is reflexive. The relation is obviously symmetric.

To show transitivity we consider a bordism $(\mathbf{W}, g)$ between $(\mathbf{S}_0, g_0)$ and $(\mathbf{S}_1, g_1)$ and a bordism $(\mathbf{W}', g')$ between $(\mathbf{S}_1, g_1)$ and $(\mathbf{S}_2, g_2)$, where $\mathbf{W}$, $\mathbf{W}'$ and all $\mathbf{S}_i$ are regular $\mathbb{Z}/2$-oriented stratifolds. We glue $\mathbf{W}$ and $\mathbf{W}'$ along $\mathbf{S}_1$ as explained in Proposition 3.1. The result is regular and $\mathbb{Z}/2$-oriented. The boundary of $\mathbf{W} \cup_{\mathbf{S}_1} \mathbf{W}'$ is $\mathbf{S}_0 \sqcup \mathbf{S}_2$. Since g and g' agree on $\mathbf{S}_1$, they induce a map $g \cup g' : \mathbf{W} \cup_{\mathbf{S}_1} \mathbf{W}' \to X$, whose restriction to $\mathbf{S}_0$ is g_0 and to $\mathbf{S}_2$ is g_2. Thus $(\mathbf{S}_0, g_0)$ and $(\mathbf{S}_2, g_2)$ are bordant, and the relation is transitive.

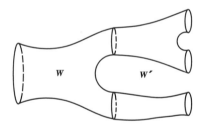

Next, we check that the equivalence classes form an abelian group with respect to the topological sum. We first note that if $(\mathbf{S}_1, g_1)$ and $(\mathbf{S}_2, g_2)$ are isomorphic, then they are bordant. A bordism is given by gluing the cylinders $(\mathbf{S}_1 \times [0,1], h)$ and $(\mathbf{S}_2 \times [0,1], h)$ via the isomorphism considered as a map from $(\{1\} \times \mathbf{S}_1)$ to $(\mathbf{S}_2 \times \{0\})$ (as explained after Proposition 3.1). Since the isomorphism classes of pairs $(\mathbf{S}, g)$ are a set and isomorphic pairs are bordant, the bordism classes are a quotient set of the isomorphism classes, and thus are a set.

The operation on $SH_m(X; \mathbb{Z}/2)$ defined by the topological sum satisfies all the axioms of an abelian group. The topological sum is associative and commutative. An element $(\mathbf{S}, g)$ represents the zero element if an only if there is a bordism $(\mathbf{T}, h)$ with $\partial(\mathbf{T}, h) = (\mathbf{S}, g)$. The inverse of $[\mathbf{S}, g]$ is given

by $[\mathbf{S}, g]$ again, since $[\mathbf{S}, g] + [\mathbf{S}, g]$ is the boundary of $(\mathbf{S} \times [0, 1], h)$, the cylinder over $(\mathbf{S}, g)$.
q.e.d.

Remark: By the last argument, each element $[\mathbf{S}, g]$ in $SH_m(X; \mathbb{Z}/2)$ is 2-torsion, i.e., $2[\mathbf{S}, g] = 0$. In other words, $SH_m(X; \mathbb{Z}/2)$ is a vector space over the field $\mathbb{Z}/2$.

Here we abbreviate the quotient group $\mathbb{Z}/2\mathbb{Z}$, which is a field, as $\mathbb{Z}/2$. Later we will define $SH_m(X; \mathbb{Q})$, which will be a $\mathbb{Q}$-vector space. This indicates the role of $\mathbb{Z}/2$ in the notation of homology groups.

To obtain a feeling for homology groups, we compute $SH_0(\text{pt}; \mathbb{Z}/2)$, the 0-th homology group of a point. A 0-dimensional stratifold is the same as a 0-dimensional manifold, and a 1-dimensional c-stratifold that is $\mathbb{Z}/2$-oriented, is the same as a 1-dimensional manifold with a germ of collars since the codimension-1 stratum is empty and there is only one possible non-empty stratum. We recall from [**Mi 2**, Appendix)] that a compact 1-dimensional manifold W with boundary has an even number of boundary points. Thus the number of points modulo 2 of a closed 0-dimensional manifold is a bordism invariant. On the other hand, an even number of points is the boundary of a disjoint union of intervals. We conclude:

Theorem 4.5. $SH_0(\text{pt}; \mathbb{Z}/2) \cong \mathbb{Z}/2$. *The isomorphism is given by the number of points modulo 2. The non-trivial element is* $[\text{pt}, \text{id}]$.

There is a generalization of Theorem 4.5; one can determine $SH_0(X; \mathbb{Z}/2)$ for an arbitrary space X. To develop this, we remind the reader of the following definition:

Definition: *A topological space X is called* **path connected** *if any two points x and y in X can be connected by a path, i.e., there is a continuous map $\alpha : [a, b] \to X$ with $\alpha(a) = x$ and $\alpha(b) = y$.*

The relation on X which says two points are equivalent if they can be joined by a path is an equivalence relation. The equivalence classes are called the **path components** of X. A path connected space is connected (why?) but the converse is in general not true (why?) although it is, for example, true for manifolds (why?).

The number of path components is an interesting invariant of a topological space. It can be computed via homology. Recall that since all elements of $SH_m(X;\mathbb{Z}/2)$ are 2-torsion, we consider $SH_m(X;\mathbb{Z}/2)$ as a $\mathbb{Z}/2$-vector-space.

Theorem 4.6. *The number of path components of a topological space X is equal to $\dim_{\mathbb{Z}/2} SH_0(X;\mathbb{Z}/2)$. A basis of $SH_0(X,\mathbb{Z}/2)$ (as $\mathbb{Z}/2$-vector space) is given by the homology classes $[\text{pt}, g_i]$, where g_i maps the point to an arbitrary point of the i-th path component of X.*

Proof: We recall that $\mathbb{Z}/2$-oriented c-stratifolds of dimension ≤ 1 are the same as manifolds with germs of collars. Choose for each such path component X_i a point x_i in X_i. Then we consider the bordism class $\alpha_i := [\text{pt}, x_i]$, where the latter means the 0-dimensional manifold pt together with the map mapping this point to x_i. We claim that the bordism classes α_i form a basis of $SH_0(X;\mathbb{Z}/2)$. This follows from the definition of path components and bordism classes once we know that for points x and y in X, we have $[\text{pt}, x] = [\text{pt}, y]$ if and only if there is a path joining x and y. If x and y can be joined by a path then the path is a bordism from (pt, x) to (pt, y) and so $[\text{pt}, x] = [\text{pt}, y]$. Conversely, if there is a bordism between $[\text{pt}, x]$ and $[\text{pt}, y]$, we consider the path components of this bordism that have a non-empty boundary. We know that each such path component is homeomorphic to $[0, 1]$ ([**Mi 2**], Appendix). Since the boundary consists of two points, there can be only one path component with non-empty boundary and this path component of the bordism is itself a path joining x and y.
q.e.d.

As one can see from the proof, this result is more or less a tautology. Nevertheless, it turns out that the interpretation of the number of path components as the dimension of $SH_0(X;\mathbb{Z}/2)$ is very useful. We will develop methods for the computation of $SH_0(X;\mathbb{Z}/2)$ which involve also higher homology groups $SH_k(X;\mathbb{Z}/2)$ for $k > 0$ and apply them, for example, to prove a sort of Jordan separation theorem in §6.

One of the main reasons for introducing stratifold homology groups is that one can use them to distinguish spaces. To compare the stratifold homology of different spaces we define **induced maps**.

Definition: *For a continuous map $f : X \to Y$, define $f_* : SH_m(X;\mathbb{Z}/2) \to SH_m(Y;\mathbb{Z}/2)$ by $f_*([\mathbf{S}, g]) := [\mathbf{S}, fg]$.*

By construction, this is a group homomorphism. The following properties are an immediate consequence of the definition.

Proposition 4.7. *Let $f : X \to Y$ and $g : Y \to Z$ be continuous maps. Then*
$$(gf)_* = g_* f_*$$
and
$$\mathrm{id}_* = \mathrm{id}.$$

One says that $SH_m(X; \mathbb{Z}/2)$ together with the induced maps f_* is a **functor** (which means that the two properties of Proposition 4.7 are fulfilled). These functorial properties imply that if $f : X \to Y$ is a homeomorphism, then $f_* : SH_m(X; \mathbb{Z}/2) \to SH_m(Y; \mathbb{Z}/2)$ is an isomorphism. The reason is that $(f^{-1})_*$ is an inverse since $(f^{-1})_* f_* = (f^{-1} f)_* = \mathrm{id}_* = \mathrm{id}$ and similarly $f_*(f^{-1})_* = \mathrm{id}$.

We earlier motivated the idea of homology by fishing for a hole using a continuous map $g : \mathbf{S} \to X$. It is plausible that a deformation of g detects the same hole as g. These deformations play an important role in homology. The precise definition of a deformation is the notion of **homotopy**.

Definition: *Two continuous maps f and f' between topological spaces X and Y are called **homotopic** if there is a continuous map $h : X \times I \to Y$ such that $h|_{X \times \{0\}} = f$ and $h|_{X \times \{1\}} = f'$. Such a map h is called a **homotopy** from f to f'.*

One should think of a homotopy as a continuous family of maps $h_t : X \to Y$, $x \mapsto h(x, t)$ joining f and f'. Homotopy defines an equivalence relation between maps which we often denote by $\simeq$. Namely, $f \simeq f$ with homotopy $h(x, t) = f(x)$. If $f \simeq f'$ via h, this implies $f' \simeq f$ via $h'(x, t) := h(x, 1 - t)$. If $f \simeq f'$ via h and $f' \simeq f''$ via h', then $f \simeq f''$ via

$$h''(x, t) := \begin{cases} h(x, 2t) & \text{for } 0 \leq t \leq 1/2 \\ h'(x, 2t - 1) & \text{for } 1/2 \leq t \leq 1. \end{cases}$$

The reader should check that this map is continuous.

The set of all continuous maps between given topological spaces is huge and hard to analyze. Often one is only interested in those properties of a map which are unchanged under deformations. This is the reason for introducing the homotopy relation.

As indicated above, $\mathbb{Z}/2$-homology cannot distinguish objects which are equal up to deformation. This is made precise in the next result which is one of the fundamental properties of homology and is given the name **homotopy axiom**:

Proposition 4.8. *Let f and f' be homotopic maps from X to Y. Then*

$$f_* = f'_* : SH_m(X; \mathbb{Z}/2) \to SH_m(Y; \mathbb{Z}/2).$$

Proof: Let $h : X \times I \to Y$ be a homotopy between maps f and f' from X to Y. Consider $[\mathbf{S}, g] \in SH_m(X)$. Then the cylinder $(\mathbf{S} \times [0,1], h \circ (g \times \mathrm{id}))$ is a bordism between $(\mathbf{S}, fg)$ and $(\mathbf{S}, f'g)$, and thus $f_*[\mathbf{S}, g] = f'_*[\mathbf{S}, g]$.
q.e.d.

We mentioned above that homeomorphisms induce isomorphisms between $SH_m(X; \mathbb{Z}/2)$ and $SH_m(Y; \mathbb{Z}/2)$. This can be generalized by introducing **homotopy equivalences**. We say that a continuous map $f : X \to Y$ is a **homotopy equivalence** if there is a continuous map $g : Y \to X$ such that gf and fg are homotopic to id_X and id_Y, the identity maps on X and Y respectively. Such a map g is called a **homotopy inverse** to f. Roughly, a homotopy equivalence is a deformation from one space to another. For example, the inclusion $i : S^m \to \mathbb{R}^{m+1} - \{0\}$ is a homotopy equivalence with homotopy inverse given by $g : x \mapsto x/||x||$. We have $gi = \mathrm{id}_{S^m}$ and $h(x, t) = tx + (1-t)\frac{x}{||x||}$ is a homotopy between ig and $\mathrm{id}_{\mathbb{R}^{m+1}-\{0\}}$. A homotopy equivalence induces an isomorphism in stratifold homology:

Proposition 4.9. *A homotopy equivalence $f : X \to Y$ induces isomorphisms $f_* : SH_k(X; \mathbb{Z}/2) \to SH_k(Y; \mathbb{Z}/2)$ for all k.*

The reason is that if g is a homotopy inverse of f then Propositions 4.7 and 4.8 ensure that g_* is an inverse of f_*.

A space is called **contractible** if it is homotopy equivalent to a point. For example, $\mathbb{R}^n$ is contractible. Thus for contractible spaces one has an isomorphism between $SH_n(X; \mathbb{Z}/2)$ and $SH_n(\mathrm{pt}; \mathbb{Z}/2)$. This gives additional motivation to determine the higher homology groups of a point. The answer is very simple:

Theorem 4.10. *For $n > 0$ we have*

$$SH_n(\mathrm{pt}; \mathbb{Z}/2) = 0.$$

Proof: Since there is only the constant map to the space consisting of a single point we can omit the maps in our bordism classes if the space X is a point. Thus we have to show that each $\mathbb{Z}/2$-oriented compact regular stratifold $\mathbf{S}$ of dimension > 0 is the boundary of a $\mathbb{Z}/2$-oriented compact regular c-stratifold $\mathbf{T}$. There is an obvious candidate, the closed cone $C\mathbf{S}$ defined in chapter 2, §2. This is obviously $\mathbb{Z}/2$-oriented since $\mathbf{S}$ is $\mathbb{Z}/2$-oriented and the dimension of $\mathbf{S}$ is > 0. (If $\dim \mathbf{S} = 0$, then the 0-dimensional stratum, the top of the cone, is not empty in a 1-dimensional stratifold, and so the cone is not $\mathbb{Z}/2$-oriented!) We have seen already that the cone is regular if $\mathbf{S}$ is regular.
q.e.d.

This is a good place to see the effect of restricting to $\mathbb{Z}/2$-oriented stratifolds. If we considered arbitrary stratifolds, then even in dimension 0 the homology group of a point would be trivial. But if all the homology groups of a point are zero, then the homology groups of any nice space are also zero. This follows from the Mayer-Vietoris sequence which we will introduce in the next chapter. Similarly, the homology groups would be uninteresting if we did not require that the stratifolds are compact since, as we already remarked, the half open cylinder $\mathbf{S} \times [0, 1)$ could be taken to show that $[\mathbf{S}]$ is zero.

5. Exercises

(1) Let $\mathbf{S}$ and $\mathbf{S}'$ be two stratifolds which are homeomorphic. Assuming that S is Z/2- orientable, does it follow that $\mathbf{S}'$ is $\mathbb{Z}/2$-orientable?

(2) Show directly that the gluing of two regular stratifolds along a common boundary is regular.

(3) Give a condition for a p-stratifold with two strata to be regular.

(4) Where did we use regularity?

(5) Show that the stratifold $\mathbf{S}$ from exercise 3 in chapter 2 is not regular. Construct a smooth function to $\mathbb{R}$ where the regular points are not open. Construct a stratifold $\mathbf{S}$ and a smooth function to $\mathbb{R}$, such that the regular values are not open. (Hint: Modify the construction of the stratifold from exercise 3 in chapter 2 to obtain a stratifold structure on the upper half plane together with

a retraction to the 1-stratum $\mathbb{R}$ whose critical values are $1/n$ for $n = 1, 2, \ldots$.)

(6) Let **S** be a compact m-dimensional $\mathbb{Z}/2$-oriented regular stratifold. Give a necessary and sufficient condition such that $[\mathbf{S}, \mathrm{id}] = 0$.

(7) Classify all compact connected oriented zero, one and two dimensional p-stratifolds up to homeomorphism.

(8) Let $f : X \to Y$ be a continuous map between two topological spaces.
a) Show that if X_α is a path component in X then $f(X_\alpha)$ is contained in a path component of Y.
b) We saw that $SH_0(X; \mathbb{Z}/2)$ and $SH_0(Y; \mathbb{Z}/2)$ are vector spaces having a basis in one-to-one correspondence to their path components. Show that $f_* : SH_0(X; \mathbb{Z}/2) \to SH_0(Y; \mathbb{Z}/2)$ maps the basis element corresponding to a path component X_α in X to the basis element corresponding to the path component in Y that contains $f(X_\alpha)$.

(9) Let $X = S^1 \times S^1$ be a torus with an embedding of the open ball B^2. Remove the open ball $\frac{1}{2}B^2$ from X. Let S denote the boundary of X, thus S is homeomorphic to S^1. Show that the inclusion of S in X is not homotopic to the constant map but it is bordant to zero, that is the inclusion map $S \to X$ represents the zero element in $SH_1(X; \mathbb{Z}/2)$.

(10) a) Show that the identity map and the antipodal map $A : S^{2n+1} \to S^{2n+1}$ are homotopic. (Hint: show this first for $n = 0$.)
b) Show that a map $f : S^n \to X$ is homotopic to a constant map if and only if it can be extended to a map from D^{n+1}.
c) A subset $U \subseteq \mathbb{R}^n$ is called star-shaped if there is a point $x \in U$ such that for every $y \in U$ the interval joining x and y is contained in U. If U is star-shaped, show that any two maps $f_i : X \to U$ are homotopic. Deduce that each star-shaped set is contractible.

(11) Show that S^n is not contractible but S^∞ is contractible ($S^\infty = \bigcup S^n$ with the topology τ such that a subset $U \subseteq S^\infty$ is open if and only if $U \cap S^n$ is open in S^n for all n).

(12) Let X be a topological space and A be a subspace. A is called a retract of X if there is a continuous map $r : X \to A$ such that $r|_A = \mathrm{id}$. Show that in this case r_* is surjective and i_* is injective where $i : A \to X$ is the inclusion. Deduce that if A is a retract of X and $SH_n(X; \mathbb{Z}/2) = \{0\}$ then $SH_n(A; \mathbb{Z}/2) = \{0\}$ and if $SH_n(A; \mathbb{Z}/2) \neq \{0\}$ then $SH_n(X; \mathbb{Z}/2) \neq \{0\}$. Show that if X is contractible and A is a retract of X then A is contractible.

5. Exercises

(13) Let X be a topological space and A a subspace. We say that A is a deformation retract of X if there is a homotopy $H : X \times I \to X$ such that the restriction of H to $A \times I$ is the projection on the first factor, and for every $x \in X$ we have $H(x,0) = x$ and $H(x,1) \in A$. Clearly, if A is a deformation retract of X then it is a retract of X.
a) Show that if A is a deformation retract of X then A is homotopy equivalent to X.
b) Denote by $h_1 = H|_{X \times \{1\}}$ and i the inclusion of A. Show that h_{1*} and i_* are isomorphisms.

(14) Show that the following spaces are homotopy equivalent:
a) $\mathbb{R}^n \setminus \{0\}$ and S^{n-1}.
b) The 2-torus after identifying one copy of S^1 to a point and $S^2 \vee S^1$ ($S^2 \vee S^1 = S^2 \sqcup S^1/(1,0,0) \sim (1,0)$).
c) $S^n \times S^m \setminus \{\text{pt}\}$ and $S^n \vee S^m$.

(15) Let M be the manifold obtained from $\mathbb{RP}^n$ ($\mathbb{CP}^n$) by removing one point. Can you find a closed manifold which is homotopy equivalent to M?

Chapter 5

The Mayer-Vietoris sequence and homology groups of spheres

1. The Mayer-Vietoris sequence

While on the one hand the definition of $SH_n(X; \mathbb{Z}/2)$ is elementary and intuitive, on the other hand it is hard to imagine how one can compute these groups. In this chapter we will describe an effective method which, in combination with the homotopy axiom (4.8), will often allow us to reduce the computation of $SH_n(X; \mathbb{Z}/2)$ to our knowledge of $SH_m(\mathrm{pt}; \mathbb{Z}/2)$. We will discuss interesting applications of these computations in the next chapter.

The method for relating $SH_n(X; \mathbb{Z}/2)$ to $SH_m(\mathrm{pt}; \mathbb{Z}/2)$ is based on Propositions 4.7, 4.8 and 4.9, and the following long exact sequence. To formulate the method, we have to introduce the notion of **exact sequences**. A sequence of homomorphisms between abelian groups

$$\cdots \to A_n \xrightarrow{f_n} A_{n-1} \xrightarrow{f_{n-1}} A_{n-2} \to \cdots$$

is called **exact** if for each n $\ker f_{n-1} = \operatorname{im} f_n$.

For example,

$$0 \to \mathbb{Z} \xrightarrow{\cdot 2} \mathbb{Z} \to \mathbb{Z}/2 \to 0$$

is exact where the map $\mathbb{Z} \to \mathbb{Z}/2$ is the reduction mod 2. The zeros on the left and right side mean in combination with exactness that the map $\mathbb{Z} \xrightarrow{\cdot 2} \mathbb{Z}$ is injective and $\mathbb{Z} \to \mathbb{Z}/2$ is surjective, which is clearly the case. The exactness in the middle means that the kernel of the reduction mod 2 is the image of the multiplication by 2, which is also clear.

To get a feeling for exact sequences, we observe that if we have an exact sequence

$$A \xrightarrow{0} B \xrightarrow{f} C \xrightarrow{0} D,$$

then f is injective (following from the 0-map on the left side) and surjective (following from the 0-map on the right side). Thus, in this situation, f is an isomorphism. Of course, if there is only a 0 on the left, then f is only injective, and if there is only a 0 on the right, then f is only surjective.

Another elementary but useful consequence of exactness concerns sequences of abelian groups where each group is a finite dimensional vector space over a field K and the maps are linear maps. If

$$0 \to A_n \to A_{n-1} \to \cdots \to A_1 \to A_0 \to 0$$

is an exact sequence of finite dimensional K-vector spaces and linear maps, then:

$$\sum_{i=0}^{n} (-1)^i \dim_K A_i = 0,$$

the alternating sum of the dimensions is 0. We leave this as an elementary exercise in linear algebra for the reader.

To formulate the method, we consider the following situation. Let U and V be open subsets of a space X. We want to relate the homology groups of U, V, $U \cap V$ and $U \cup V$. To do so, we need maps between the homology groups of these spaces. There are some obvious maps induced by the different inclusions. In addition, we need a less obvious map, the so-called boundary operator $d: SH_m(U \cup V; \mathbb{Z}/2) \to SH_{m-1}(U \cap V; \mathbb{Z}/2)$. We begin with its description. Consider an element $[\mathbf{S}, g] \in SH_m(U \cup V; \mathbb{Z}/2)$. We note that $A := g^{-1}(X - V)$ and $B := g^{-1}(X - U)$ are disjoint closed subsets of $\mathbf{S}$.

1. The Mayer-Vietoris sequence

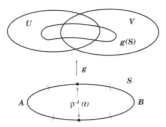

By arguments similar to the proof of Proposition 2.8, there is a separating stratifold $\mathbf{S}' \subset \mathbf{S} - (A \cup B)$ of dimension $m-1$ (the picture above explains the idea of the proof of 2.8, where $\mathbf{S}' = \rho^{-1}(t)$ for a smooth function $\rho : \mathbf{S} \to \mathbb{R}$ with $\rho(A) = 1$ and $\rho(B) = -1$ and t a regular value) and we define

$$d([\mathbf{S}, g]) := [\mathbf{S}', g|_{\mathbf{S}'}].$$

We will show in Appendix B (the proof is purely technical and plays no essential role in understanding homology) that this construction gives a well-defined map

$$d : SH_m(U \cup V; \mathbb{Z}/2) \to SH_{m-1}(U \cap V; \mathbb{Z}/2).$$

If we apply this construction to a topological sum, it leads to the topological sum of the corresponding pairs and so this map is a homomorphism.

Proposition 5.1. *The construction above assigning to $(\mathbf{S}, g)$ the pair $(\mathbf{S}', g|_{\mathbf{S}'})$ gives a well defined homomorphism*

$$d : SH_m(U \cup V; \mathbb{Z}/2) \to SH_{m-1}(U \cap V; \mathbb{Z}/2).$$

This map is called the **boundary operator**.

Now we can give the fundamental tool for relating the homology groups of a space X to those of a point:

Theorem 5.2. *For open subsets U and V of X the following sequence (**Mayer-Vietoris sequence**) is exact:*

$$\cdots SH_n(U \cap V; \mathbb{Z}/2) \to SH_n(U; \mathbb{Z}/2) \oplus SH_n(V; \mathbb{Z}/2) \to SH_n(U \cup V; \mathbb{Z}/2)$$

$$\xrightarrow{d} SH_{n-1}(U \cap V; \mathbb{Z}/2) \to SH_{n-1}(U; \mathbb{Z}/2) \oplus SH_{n-1}(V; \mathbb{Z}/2) \to \cdots$$

It commutes with induced maps.

Here the map $SH_n(U \cap V; \mathbb{Z}/2) \to SH_n(U; \mathbb{Z}/2) \oplus SH_n(V; \mathbb{Z}/2)$ is $\alpha \mapsto ((i_U)_*(\alpha), (i_V)_*(\alpha))$ and the map $SH_n(U; \mathbb{Z}/2) \oplus SH_n(V; \mathbb{Z}/2) \to SH_n(U \cup V; \mathbb{Z}/2)$ is $(\alpha, \beta) \mapsto (j_U)_*(\alpha) - (j_V)_*(\beta)$.

We give some explanation. The maps i_U and i_V are the inclusions from $U \cap V$ to U and V, the maps j_U and j_V are the inclusions from U and V to $U \cup V$. The sequence extends arbitrarily far to the left and ends on the right with

$$\cdots \to SH_0(U; \mathbb{Z}/2) \oplus SH_0(V; \mathbb{Z}/2) \to SH_0(U \cup V; \mathbb{Z}/2) \to 0.$$

Finally, the last condition in the theorem means that if we have a space X' with open subspaces U' and V' and a continuous map $f : X \to X'$ with $f(U) \subset U'$ and $f(V) \subset V'$, then the diagram

$$\begin{array}{ccc}
\cdots \to SH_n(U \cap V; \mathbb{Z}/2) & \to & SH_n(U; \mathbb{Z}/2) \oplus SH_n(V; \mathbb{Z}/2) \to \\
\downarrow (f|_{U \cap V})_* & & \downarrow (f|_U)_* \oplus (f|_V)_* \\
\cdots \to SH_n(U' \cap V'; \mathbb{Z}/2) & \to & SH_n(U'; \mathbb{Z}/2) \oplus SH_n(V'; \mathbb{Z}/2) \to \\
\to SH_n(U \cup V; \mathbb{Z}/2) & \xrightarrow{d} & SH_{n-1}(U \cap V; \mathbb{Z}/2) \to \cdots \\
\downarrow (f|_{U \cup V})_* & & \downarrow (f|_{U \cap V})_* \\
\to SH_n(U' \cup V'; \mathbb{Z}/2) & \xrightarrow{d} & SH_{n-1}(U' \cap V'; \mathbb{Z}/2) \to \cdots
\end{array}$$

commutes. That is to say that the two compositions of maps going from the upper left corner to the lower right corner in any rectangle agree.

The reader might wonder why we have taken the difference map $(j_U)_*(\alpha) - (j_V)_*(\beta)$ instead of the sum $(j_U)_*(\alpha) + (j_V)_*(\beta)$, which is equivalent in our situation since for all homology classes $\alpha \in SH_m(X; \mathbb{Z}/2)$ we have $\alpha = -\alpha$. The reason is that a similar sequence exists for other homology groups (the existence of the Mayer-Vietoris sequence is actually one of the basic axioms for a homology theory as will be explained later) where the elements do not have order 2, and thus one has to take the difference map to obtain an exact sequence. We will give the proof in such a way that it will extend verbatim to the other main homology groups in this book—integral homology—so that we don't have to repeat the argument.

The idea of the proof of Theorem 5.2 is very intuitive but there are some technical points which make it rather lengthy. We now give a short proof explaining the fundamental steps. Understanding this short proof is very helpful for getting a general feeling for homology theories. In Appendix B we add the necessary details which may be skipped in a first reading of the book.

1. The Mayer-Vietoris sequence

Short proof of Theorem 5.2: We will show the exactness of the Mayer-Vietoris sequence step by step. We first recall that a sequence

$$A \xrightarrow{f} B \xrightarrow{g} C$$

is exact if and only if $gf = 0$ (i.e., $\text{im } f \subset \ker g$) and $\ker g \subset \text{im } f$.

We first consider the sequence

$$SH_n(U \cap V; \mathbb{Z}/2) \to SH_n(U; \mathbb{Z}/2) \oplus SH_n(V; \mathbb{Z}/2) \to SH_n(U \cup V; \mathbb{Z}/2).$$

Obviously the composition of the two maps is zero. To show the other inclusion, we consider $[\mathbf{S}, g] \in SH_n(U; \mathbb{Z}/2)$ and $[\mathbf{S}', g'] \in SH_n(V; \mathbb{Z}/2)$ such that $([\mathbf{S}, g], [\mathbf{S}', g'])$ maps to zero in $SH_n(U \cup V; \mathbb{Z}/2)$. Let $(\mathbf{T}, h)$ be a zero bordism of $[\mathbf{S}, j_U g] - [\mathbf{S}', j_V g']$. Then we separate $\mathbf{T}$ using Proposition 2.8 along a compact regular stratifold $\mathbf{D}$ with $h(\mathbf{D}) \subset U \cap V$. We will show in the detailed proof that we actually can choose $\mathbf{T}$ such that there is an open neighbourhood U of $\mathbf{D}$ in $\mathbf{T}$ and an isomorphism of $\mathbf{D} \times (-\epsilon, \epsilon)$ to U, which on $\mathbf{D} \times \{0\}$ is the identity map. In other words: a bicollar exists (this is where we apply the property that homology classes consist of regular stratifolds). Then, as explained in §4, we can cut along $\mathbf{D}$ to obtain a bordism $(\mathbf{T}_-, h|_{\mathbf{T}_-})$ between $(\mathbf{S}, g)$ and $(\mathbf{D}, h|_{\mathbf{D}})$ as well as a bordism $(\mathbf{T}_+, g|_{\mathbf{T}_+})$ between $(\mathbf{D}, h|_{\mathbf{D}})$ and $(\mathbf{S}', g')$. Thus $[\mathbf{D}, h|_{\mathbf{D}}] \in SH_n(U \cap V; \mathbb{Z}/2)$ maps to $([\mathbf{S}, g], [\mathbf{S}', g']) \in SH_n(U; \mathbb{Z}/2) \oplus SH_n(V; \mathbb{Z}/2)$.

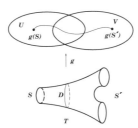

Next we consider the exactness of

$$SH_n(U \cup V; \mathbb{Z}/2) \xrightarrow{d} SH_{n-1}(U \cap V; \mathbb{Z}/2) \to SH_{n-1}(U; \mathbb{Z}/2) \oplus SH_{n-1}(V; \mathbb{Z}/2).$$

The composition of the two maps is zero. For this we show in the detailed proof that, as above, we can choose a representative for the homology class in $U \cup V$ such that we can cut along the separating stratifold defining the boundary operator. The argument is demonstrated in the following figure.

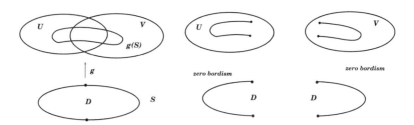

The other inclusion is demonstrated by the same pictures read in reverse order, where instead of cutting we glue.

Finally, we prove the exactness of

$$SH_n(U;\mathbb{Z}/2) \oplus SH_n(V;\mathbb{Z}/2) \to SH_n(U \cup V;\mathbb{Z}/2) \xrightarrow{d} SH_{n-1}(U \cap V;\mathbb{Z}/2).$$

If $[\mathbf{S},g] \in SH_n(U;\mathbb{Z}/2)$, we show $d(j_U)_*[\mathbf{S},g] = 0$. This is obvious by the construction of the boundary operator since we can choose ρ and the regular value t such that the separating regular stratifold $\mathbf{D}$ is empty. By the same argument $d(j_V)_*$ is the trivial map.

To show the other inclusion we consider $[\mathbf{S},g] \in SH_n(U \cup V;\mathbb{Z}/2)$ with $d([\mathbf{S},g]) = 0$. We will show in Appendix B that we can choose $(\mathbf{S},g)$ in such a way that the regular stratifold $\mathbf{S}$ is obtained from two regular c-stratifolds $\mathbf{S}_+$ and $\mathbf{S}_-$ with same boundary $\mathbf{D}$ by gluing them together along $\mathbf{D}$. Furthermore, we have $g(\mathbf{S}_+) \subset U$ and $g(\mathbf{S}_-) \subset V$.

If $d([\mathbf{S},g]) = 0$, there is a compact regular c-stratifold $\mathbf{Z}$ with $\partial \mathbf{Z} = \mathbf{D}$ and an extension of $g|_\mathbf{D}$ to $r: \mathbf{Z} \to U \cap V$. We glue $\mathbf{S}_+$ and $\mathbf{S}_-$ to $\mathbf{Z}$ to obtain $\mathbf{S}_+ \cup_\mathbf{D} \mathbf{Z}$ and $\mathbf{S}_- \cup_\mathbf{D} \mathbf{Z}$, and map the first to U via $g|_{\mathbf{S}_+} \cup r$ and the second to V via $g|_{\mathbf{S}_-} \cup r$. This gives an element of $SH_n(U;\mathbb{Z}/2) \oplus SH_n(V;\mathbb{Z}/2)$.

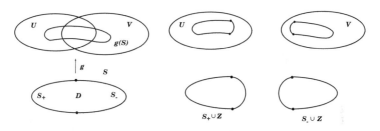

We are finished if in $U \cup V$ the difference of these two bordism classes is equal to $[\mathbf{S},g]$. For this we take the c-stratifolds $(\mathbf{S}_+ \cup_\mathbf{D} \mathbf{Z}) \times [0,1]$ and $(\mathbf{S}_- \cup_\mathbf{D} \mathbf{Z}) \times [1,2]$ and paste them together along $\mathbf{Z} \times 1$.

2. Reduced homology groups and homology groups of spheres

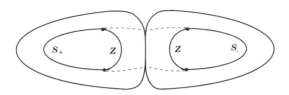

We will show in Appendix A that this stratifold can be given the structure of a regular c-stratifold with boundary $\mathbf{S_+} \cup_\mathbf{D} \mathbf{Z} + \mathbf{S_-} \cup_\mathbf{D} \mathbf{Z} + \mathbf{S_+} \cup_\mathbf{D} \mathbf{S_-}$. Since $\mathbf{S_+} \cup_\mathbf{D} \mathbf{S_-} = \mathbf{S}$ and our maps extend to a map from this regular c-stratifold to $U \cup V$, we have a bordism between $[\mathbf{S_+} \cup_\mathbf{D} \mathbf{Z}, g|_{\mathbf{S_+}} \cup r] + [\mathbf{S_-} \cup_\mathbf{D} \mathbf{Z}, g|_{M_-} \cup r]$ and $[\mathbf{S}, g]$.
q. e. d.

As an application we compute the homology groups of a topological sum. Let X and Y be topological spaces and $X \sqcup Y$ the topological sum (the disjoint union). Then X and Y are open subspaces of $X \sqcup Y$ and we denote them by U and V. Since the intersection $U \cap V$ is the empty set and the homology groups of the empty set are 0 (this is a place where it is necessary to allow the empty set as k-dimensional stratifold whose corresponding homology groups are of course 0) the Mayer-Vietoris sequence gives short exact sequences:

$$0 \to SH_k(X; \mathbb{Z}/2) \oplus SH_k(Y; \mathbb{Z}/2) \to SH_k(X \sqcup Y; \mathbb{Z}/2) \to 0,$$

where the zeroes on the left and right side correspond to $SH_n(\emptyset; \mathbb{Z}/2) = 0$ and $SH_{n-1}(\emptyset; \mathbb{Z}/2) = 0$, respectively. The map in the middle is $(j_X)_* - (j_Y)_*$. As explained above, exactness implies that this map is an isomorphism:

$$(j_X)_* - (j_Y)_* : SH_n(X; \mathbb{Z}/2) \oplus SH_n(Y; \mathbb{Z}/2) \to SH_n(X \sqcup Y; \mathbb{Z}/2).$$

Of course, this also implies that the sum $(j_X)_* + (j_Y)_*$ is an isomorphism.

2. Reduced homology groups and homology groups of spheres

For computations it is often easier to split the homology groups into the homology groups of a point and the "rest", which will be called reduced homology. Let $p : X \to \mathrm{pt}$ be the constant map to the space consisting of a single point. The n-th **reduced homology group** is $\widetilde{SH}_n(X; \mathbb{Z}/2) := \ker(p_* : SH_n(X; \mathbb{Z}/2) \to SH_n(\mathrm{pt}; \mathbb{Z}/2))$. A continuous map $f : X \to Y$ induces a homomorphism on the reduced homology groups by restriction to the kernels and we denote it again by $f_* : \widetilde{SH}_n(X; \mathbb{Z}/2) \to \widetilde{SH}_n(Y; \mathbb{Z}/2)$.

If X is non-empty, there is a simple relation between the homology and the reduced homology of X:

$$SH_n(X; \mathbb{Z}/2) \cong \widetilde{SH}_n(X; \mathbb{Z}/2) \oplus SH_n(\text{pt}; \mathbb{Z}/2).$$

The isomorphism sends a homology class $a \in SH_n(X; \mathbb{Z}/2)$ to the pair $(a - i_*p_*(a), p_*a)$, where i is the inclusion from pt to an arbitrary point in X. For $n > 0$ this means that the reduced homology is the same as the unreduced homology, but for $n = 0$ it differs by a summand $\mathbb{Z}/2$.

Since it is often useful to work with reduced homology, it would be nice to know if there is also a Mayer-Vietoris sequence for reduced homology. This is the case. We prepare for the argument by developing a useful algebraic result. Consider a commutative diagram of abelian groups and homomorphisms

$$\begin{array}{ccccccc} A_1 & \xrightarrow{f_1} & A_2 & \xrightarrow{f_2} & A_3 & \xrightarrow{f_3} & A_4 \\ \downarrow h_1 & & \downarrow h_2 & & \downarrow h_3 & & \downarrow h_4 \\ B_1 & \xrightarrow{g_1} & B_2 & \xrightarrow{g_2} & B_3 & \xrightarrow{g_3} & B_4 \end{array}$$

where the horizontal sequences are exact and the map h_1 is surjective. Then we consider the sequence

$$\ker h_1 \xrightarrow{f_1|} \ker h_2 \xrightarrow{f_2|} \ker h_3 \xrightarrow{f_3|} \ker h_4$$

where the maps $f_i|$ are $f_i|_{\ker h_i}$. The statement is that the sequence

$$\ker h_2 \xrightarrow{f_2|} \ker h_3 \xrightarrow{f_3|} \ker h_4$$

is again exact. This is proved by a general method called **diagram chasing**, which we introduce in proving this statement. We chase in the commutative diagram given by A_i and B_j above. The first step is to show that im $f_2| \subset \ker f_3|$ or equivalently $(f_3|)(f_2|) = 0$. This follows since $f_3 f_2 = 0$. To show that ker $f_3| \subset$ im $f_2|$, we start the chasing by considering $x \in \ker h_3$ with $f_3(x) = 0$. By exactness of the sequence given by the A_i, there is $y \in A_2$ with $f_2(y) = x$. Since $h_3(x) = 0$ and $h_3 f_2(y) = g_2 h_2(y)$, we have $g_2 h_2(y) = 0$ and thus by the exactness of the lower sequence and the surjectivity of h_1, there is $z \in A_1$ with $g_1 h_1(z) = h_2(y)$. Since $g_1 h_1(z) = h_2 f_1(z)$, we conclude $h_2(y - f_1(z)) = 0$ or $y - f_1(z) \in \ker h_2$. Since $f_2 f_1(z) = 0$, we are done since we have found $y - f_1(z) \in \ker h_2$ with $f_2(y - f_1(z)) = f_2(y) = x$.

With this algebraic information, we can compare the Mayer-Vietoris sequences for $X = U \cup V$ with that of the space pt given by $U' := \text{pt} =: V'$:

2. Reduced homology groups and homology groups of spheres

$$\cdots \to SH_n(U \cap V; \mathbb{Z}/2) \to SH_n(U; \mathbb{Z}/2) \oplus SH_n(V; \mathbb{Z}/2) \to$$
$$\downarrow \qquad\qquad \downarrow$$
$$\to SH_n(U' \cap V'; \mathbb{Z}/2) \to SH_n(U'; \mathbb{Z}/2) \oplus SH_n(V'; \mathbb{Z}/2) \to$$

$$\to SH_n(X; \mathbb{Z}/2) \qquad \to \qquad SH_{n-1}(U \cap V; \mathbb{Z}/2)$$
$$\downarrow \qquad\qquad\qquad\qquad \downarrow$$
$$\to SH_n(U' \cup V'; \mathbb{Z}/2) \to \qquad SH_{n-1}(U' \cap V'; \mathbb{Z}/2) \to \cdots.$$

Since $U' \cap V' = U' = V' = U' \cup V' = \text{pt}$, all vertical maps are surjective if $U \cap V$ is non-empty (and thus U and V as well), and therefore by the argument above, the reduced Mayer-Vietoris sequence

$$\cdots \to \widetilde{SH}_n(U \cap V; \mathbb{Z}/2) \to \widetilde{SH}_n(U; \mathbb{Z}/2) \oplus \widetilde{SH}_n(V; \mathbb{Z}/2)$$
$$\to \widetilde{SH}_n(X; \mathbb{Z}/2) \to \widetilde{SH}_{n-1}(U \cap V; \mathbb{Z}/2) \to \cdots$$

is exact if $U \cap V$ is non-empty.

Now we use the homotopy axiom (Proposition 4.8) and the reduced Mayer-Vietoris sequence to express the homology of the sphere $S^m := \{x \in \mathbb{R}^{m+1} \mid \|x\| = 1\}$ in terms of the homology of a point. For this we decompose S^m into the complement of the north pole $N = (0, ..., 0, 1)$ and the south pole $S = (0, ..., 0, -1)$, and define $S_+^m : S^m - \{S\}$ and $S_-^m := S^m - \{N\}$. The inclusion $S^{m-1} \to S_+^m \cap S_-^m$ mapping $y \mapsto (y, 0)$ is a homotopy equivalence with homotopy inverse $r : (x_1, \ldots, x_{m+1}) \mapsto (x_1, \ldots, x_m)/\|(x_1,\ldots,x_m)\|$ (why?). Both S_+^m and S_-^m are homotopy equivalent to a point, or equivalently the identity map on these spaces is homotopic to the constant map (why?). Since $S_+^m \cup S_-^m$ is S^m, the reduced Mayer-Vietoris sequence gives an exact sequence

$$\cdots \to \widetilde{SH}_n(S_+^m \cap S_-^m; \mathbb{Z}/2) \to \widetilde{SH}_n(S_+^m; \mathbb{Z}/2) \oplus \widetilde{SH}_n(S_-^m; \mathbb{Z}/2)$$
$$\to \widetilde{SH}_n(S^m; \mathbb{Z}/2) \xrightarrow{d} \widetilde{SH}_{n-1}(S_+^m \cap S_-^m; \mathbb{Z}/2) \to \cdots.$$

If we use the isomorphisms induced by the homotopy equivalences above this becomes

$$\cdots \to \widetilde{SH}_n(S^{m-1}; \mathbb{Z}/2) \to \widetilde{SH}_n(\text{pt}; \mathbb{Z}/2) \oplus \widetilde{SH}_n(\text{pt}; \mathbb{Z}/2)$$
$$\to \widetilde{SH}_n(S^m; \mathbb{Z}/2) \xrightarrow{d} \widetilde{SH}_{n-1}(S^{m-1}; \mathbb{Z}/2) \to \cdots.$$

Since $\widetilde{SH}_k(\text{pt}; \mathbb{Z}/2) = 0$, we obtain an isomorphism

$$d : \widetilde{SH}_n(S^m; \mathbb{Z}/2) \xrightarrow{\cong} \widetilde{SH}_{n-1}(S^{m-1}; \mathbb{Z}/2)$$

and so by induction:
$$\widetilde{SH}_n(S^m; \mathbb{Z}/2) \xrightarrow{\cong} \widetilde{SH}_{n-m}(S^0; \mathbb{Z}/2).$$
The space S^0 consists of two points $\{+1\}$ and $\{-1\}$ which are open subsets and by the formula above for the homology of a topological sum we have $\widetilde{SH}_n(S^0; \mathbb{Z}/2) \cong SH_n(\text{pt}; \mathbb{Z}/2)$. We summarize:

Theorem 5.3.
$$\widetilde{SH}_n(S^m; \mathbb{Z}/2) \cong SH_{n-m}(\text{pt}; \mathbb{Z}/2)$$
or
$$SH_n(S^m; \mathbb{Z}/2) \cong SH_n(\text{pt}; \mathbb{Z}/2) \oplus SH_{n-m}(\text{pt}; \mathbb{Z}/2).$$
In particular, for $m > 0$ we have for $k = 0$ or $k = m$
$$SH_k(S^m; \mathbb{Z}/2) = \mathbb{Z}/2$$
and
$$SH_k(S^m; \mathbb{Z}/2) = 0$$
otherwise.

It is natural to ask for an explicit representative of the non-trivial element in $SH_m(S^m; \mathbb{Z}/2)$ for $m > 0$. For this we introduce the fundamental class of a compact $\mathbb{Z}/2$-oriented regular stratifold. Let **S** be a n-dimensional $\mathbb{Z}/2$-oriented compact regular stratifold. We define its **fundamental class** as $[\mathbf{S}]_{\mathbb{Z}/2} := [\mathbf{S}, \text{id}] \in SH_n(\mathbf{S}, \mathbb{Z}/2)$. As the name indicates this class is important. We will see that it is always non-zero. In particular, we obtain for each compact smooth manifold the fundamental class $[M]_{\mathbb{Z}/2} = [M, \text{id}]$. In the case of the spheres the non-vanishing is clear since by the inductive computation one sees that the non-trivial element of $SH_m(S^m; \mathbb{Z}/2)$ is given by the fundamental class $[S^m]_{\mathbb{Z}/2}$.

As an immediate consequence of Theorem 5.3 the spheres S^n and S^m are not homotopy equivalent for $m \neq n$, for otherwise their homology groups would all be isomorphic. In particular, for $n \neq m$ the spheres are not homeomorphic. In the next chapter, we will show for arbitrary manifolds that the dimension is a homeomorphism invariant.

3. Exercises

(1) Let $0 \to A \xrightarrow{f} B \xrightarrow{g} C \to 0$ be an exact sequence of abelian groups.
   a) Show the the following are equivalent:
   i) There is a map $C \xrightarrow{s} B$ such that $g \circ s = \text{id}_C$.
   ii) There is a map $B \xrightarrow{p} A$ such that $p \circ f = \text{id}_A$.
   iii) There is an isomorphism $B \xrightarrow{h} A \oplus C$ such that $h \circ f(a) = (a, 0)$

and $g \circ h^{-1}(a, c) = c$.
In this case we say that this is a split exact sequence.
b) Show that if all groups are vector spaces and all maps are linear then the sequence splits.
c) Show that if C is free then the sequence splits.

(2) Prove the five lemma: Assume that the following diagram is commutative with exact rows:

$$\begin{array}{ccccccccc} A_1 & \to & A_2 & \to & A_3 & \to & A_4 & \to & A_5 \\ \downarrow f_1 & & \downarrow f_2 & & \downarrow f_3 & & \downarrow f_4 & & \downarrow f_5 \\ B_1 & \to & B_2 & \to & B_3 & \to & B_4 & \to & B_5 \end{array}$$

a) Show that if f_2, f_4 are injective and f_1 is surjective then f_3 is injective.
b) Show that if f_2, f_4 are surjective and f_5 is injective then f_3 is surjective.
Conclude that if f_2, f_4 are isomorphisms, f_1 is surjective and f_5 is injective then f_3 is an isomorphism.
Remark: This lemma is usually used when f_1, f_2, f_4, f_5 are isomorphisms.

(3) Let $0 \to A_n \to A_{n-1} \to \cdots \to A_0 \to 0$ be an exact sequence of vector spaces and linear maps. Show that $\sum_{k=0}^{n} (-1)^k \dim(A_k) = 0$.

(4) Let $0 \to \mathbb{Z}/2 \to A \to \mathbb{Z}/2 \to 0$ be an exact sequence of abelian groups. What are the possibilities for the group A? What can you say if you know that the maps are linear maps of $\mathbb{Z}/2$ vector spaces?

(5) Compute $SH_n(X; \mathbb{Z}/2)$ for the following spaces using the Mayer-Vietoris sequence. If possible represent each class by a map from a stratifold.

a) The wedge of two circles $S^1 \vee S^1$ or more generally the wedge of two pointed spaces (X, x_0) and (Y, y_0). Assume that both x_0 and y_0 have a contractible neighborhood.
Remark: A pointed space (X, x_0) is a topological space X together with a distinguished point $x_0 \in X$. The wedge of two pointed spaces (X, x_0) and (Y, y_0) is the pointed space $(X \sqcup Y / x_0 \sim y_0, [x_0])$ and denoted by $X \vee Y$.

b) The two dimensional torus $T^2 = S^1 \times S^1$ or more generally the n-torus which is the product of n copies of S^1.

c) The Möbius band which is the space obtained from $I \times I$ after identifying the points $(0, x)$ and $(1, 1 - x)$ for every $x \in I$, where $I = [0, 1]$.

d) Let M_1 and M_2 be two smooth n-dimensional manifolds. Compute the homology of $M_1 \# M_2$ as defined in the exercises in ch. 1.

e) Any compact surface (using that an orientable surface is either homeomorphic to S^2 or a connected sum of copies of $S^1 \times S^1$, and a non-orientable surface is homeomorphic to a connected sum of copies of $\mathbb{RP}^2$).

f) Any compact orientable surface with one point removed.

g) $\mathbb{R}^2$ with n points removed. What about if we remove an infinite discrete set?

(6) Let X be a topological space.

a) Compute $SH_n(X \times S^1; \mathbb{Z}/2)$ or more generally $SH_n(X \times S^k; \mathbb{Z}/2)$ in terms of $SH_m(X; \mathbb{Z}/2)$.

b) Compute $\widetilde{SH}_n(\Sigma X; \mathbb{Z}/2)$ (the suspension of X) or more generally $\widetilde{SH}_n(\Sigma^k X; \mathbb{Z}/2)$ (the kth iterated suspension of X) in terms of $\widetilde{SH}_n(X; \mathbb{Z}/2)$, where the suspension is defined in the exercises in chapter 2.

c) Compute $SH_n(TX; \mathbb{Z}/2)$ (the suspension of X modulo both end points) or more generally $SH_n(T^k X; \mathbb{Z}/2)$ (where we define $T^k X$ by induction, $T^k X = T(T^{k-1} X)$) in terms of $SH_m(X; \mathbb{Z}/2)$.

(7) Let M be a non-empty connected closed n-dimensional manifold. Construct a map $f : M \to S^n$ with
$$f_* : SH_n(M; \mathbb{Z}/2) \to SH_n(S^n; \mathbb{Z}/2)$$
non-trivial. (Hint: Use the fact that S^n is homeomorphic to D^n with the boundary collapsed to a point.)

(8) Determine the map $SH_n(S^n; \mathbb{Z}/2) \to SH_n(\mathbb{RP}^n; \mathbb{Z}/2)$ induced by the quotient map $S^n \to \mathbb{RP}^n$.

Chapter 6

Brouwer's fixed point theorem, separation, invariance of dimension

Prerequisites: The only new ingredient used in this chapter is the definition of topological manifolds which can be found in the first pages of either [**B-J**] or [**Hi**].

1. Brouwer's fixed point theorem

Let $D^n := \{x \in \mathbb{R}^n |\ ||x|| \leq 1\}$ be the **closed unit ball** and $B^n := \{x \in \mathbb{R}^n |\ ||x|| < 1\}$ be the **open unit ball**.

Theorem 6.1. (Brouwer) *A continuous map $f : D^n \to D^n$ has a fixed point, i.e., there is a point $x \in D^n$ with $f(x) = x$.*

Proof: The case $n = 0$ is clear and so we assume that $n > 0$. If there is a continuous map $f : D^n \to D^n$ without fixed points, define $g : D^n \to S^{n-1}$ by mapping $x \in D^n$ to the intersection of the ray from $f(x)$ to x with S^{n-1} (give a formula for this map and see that it is continuous, Exercise 1).

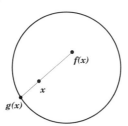

67

Then $g|_{S^{n-1}} = \text{id}_{S^{n-1}}$, the identity on S^{n-1}.

Now let $i : S^{n-1} \to D^n$ be the inclusion and consider
$$\text{id} = \text{id}_* = (g \circ i)_* = g_* \circ i_*,$$
a map
$$\widetilde{SH}_{n-1}(S^{n-1}; \mathbb{Z}/2) \xrightarrow{i_*} \widetilde{SH}_{n-1}(D^n; \mathbb{Z}/2) \xrightarrow{g_*} \widetilde{SH}_{n-1}(S^{n-1}; \mathbb{Z}/2).$$

By Theorem 5.3 we have $\widetilde{SH}_{n-1}(S^{n-1}; \mathbb{Z}/2) = \mathbb{Z}/2$. Thus the identity on $\widetilde{SH}_{n-1}(S^{n-1}; \mathbb{Z}/2)$ is non-trivial. On the other hand, since D^n is homotopy equivalent to a point, $\widetilde{SH}_{n-1}(D^n; \mathbb{Z}/2) \cong \widetilde{SH}_{n-1}(\text{pt}; \mathbb{Z}/2) = \{0\}$, implying a contradiction.

q.e.d.

2. A separation theorem

As an application of the relation between the number of path components of a space X and the dimension of $SH_0(X; \mathbb{Z}/2)$, we prove a theorem which generalizes a special case of the Jordan curve theorem (see, e.g., [**Mu**]). A topological manifold M is called closed if it is compact and has no boundary.

Theorem 6.2. *Let M be a closed, path connected, topological manifold of dimension $n - 1$ and let $f : M \times (-\epsilon, \epsilon) \to U \subset \mathbb{R}^n$ be a homeomorphism onto an open subset U of $\mathbb{R}^n$. Then $\mathbb{R}^n - f(M)$ has two path components.*

In other words, a nicely embedded closed topological manifold M of dimension $n - 1$ in $\mathbb{R}^n$ separates $\mathbb{R}^n$ into two connected components. Here "nicely embedded" means that the embedding can be extended to an embedding of $M \times (-\epsilon, \epsilon)$. If M is a smooth submanifold, then it is automatically nicely embedded [**B-J**].

Proof: Denote $\mathbb{R}^n - f(M)$ by V. Since $U \cup V = \mathbb{R}^n$ and $SH_1(\mathbb{R}^n; \mathbb{Z}/2) \cong SH_1(\text{pt}; \mathbb{Z}/2) = 0$ ($\mathbb{R}^n$ is contractible), the Mayer-Vietoris sequence gives the exact sequence
$$0 \to SH_0(U \cap V; \mathbb{Z}/2) \to SH_0(U; \mathbb{Z}/2) \oplus SH_0(V; \mathbb{Z}/2) \to SH_0(\mathbb{R}^n; \mathbb{Z}/2) \to 0.$$

Now, $U \cap V$ is homeomorphic to $M \times (-\epsilon, \epsilon) - M \times \{0\}$ and thus has two path components. Then Theorem 4.6 implies $SH_0(U \cap V; \mathbb{Z}/2)$ is 2-dimensional. The space U is homeomorphic to $M \times (-\epsilon, \epsilon)$ which is path connected implying that the dimension of $SH_0(U)$ is 1. Since the alternating sum of the dimensions is 0 we conclude $\dim_{\mathbb{Z}/2} SH_0(V; \mathbb{Z}/2) = 2$, which by Theorem

4.6 implies the statement of Theorem 6.2.
q.e.d.

As announced earlier, although the result is equivalent to a statement about $SH_0(\mathbb{R}^n - f(M); \mathbb{Z}/2)$, the proof uses higher homology groups, namely the vanishing of $SH_1(\mathbb{R}^n; \mathbb{Z}/2)$.

3. Invariance of dimension

An m-dimensional topological manifold M is a space which is locally homeomorphic to an open subset of $\mathbb{R}^m$. This is the case if and only if all points $m \in M$ have an open neighbourhood $U \ni m$ which is homeomorphic to $\mathbb{R}^m$. A fundamental question which arises immediately is whether the dimension of a topological manifold is a topological invariant. That is, could it be that M is both an m-dimensional manifold and an n-dimensional manifold for $n \neq m$? In particular, are there n and m with $n \neq m$ but with $\mathbb{R}^m \cong \mathbb{R}^m$? In this section we answer both these questions in the negative.

The key idea which we use is the local homology of a space. To define the local homology of a topological space X at a point $x \in X$, we consider the space $X \cup_{X-x} C(X-x)$, the union of X and the cone over $X - x$, where $CY = Y \times [0,1]/_{Y \times \{0\}}$ and we identify $Y \times \{1\}$ in CY with Y. Observe that $X \cup_{X-x} C(X-x) = C(X) - (x \times (0,1))$. We define the **local homology** of X at x as $SH_k(X \cup C(X-x); \mathbb{Z}/2)$. We will use the local homology of a topological manifold to characterize its dimension. For this, we need the following consideration.

Lemma 6.3. *Let M be a non-empty m-dimensional topological manifold. Then for each $x \in M$ we have*

$$\widetilde{SH}_k(M \cup C(M-x); \mathbb{Z}/2) \cong \begin{cases} \mathbb{Z}/2 & k = m \\ 0 & otherwise. \end{cases}$$

Proof: Since M is non-empty, there is an $x \in M$ and so we choose a homeomorphism φ from the open ball B^m to an open neighborhood of x. We apply the Mayer-Vietoris sequence and decompose $M \cup C(M-x)$ into $U := C(M-x)$ and $V := \varphi(B^m - \{0\}) \times (\frac{1}{2}, 1] \cup \{x\}$. The projection of V to $\varphi(B^m)$ is a homotopy equivalence and so V is contractible. Also U is contractible, since it is a cone. $U \cap V$ is homotopy equivalent (again via the projection) to $\varphi(B^m - \{0\})$ and so $U \cap V$ is homotopy equivalent to S^{m-1}. The reduced Mayer-Vietoris sequence is

$$\cdots \to \widetilde{SH}_k(U; \mathbb{Z}/2) \oplus \widetilde{SH}_k(V; \mathbb{Z}/2) \to \widetilde{SH}_k(M \cup C(M-x); \mathbb{Z}/2)$$
$$\to \widetilde{SH}_{k-1}(U \cap V; \mathbb{Z}/2) \to \cdots.$$

Since $\widetilde{SH}_k(U;\mathbb{Z}/2)$ and $\widetilde{SH}_k(V;\mathbb{Z}/2)$ are zero and $\widetilde{SH}_{k-1}(U \cap V;\mathbb{Z}/2) \cong \widetilde{SH}_{k-1}(S^{m-1};\mathbb{Z}/2)$, we have an isomorphism
$$\widetilde{SH}_k(M \cup C(M-x);\mathbb{Z}/2) \cong \widetilde{SH}_{k-1}(S^{m-1};\mathbb{Z}/2)$$
and the statement follows from 5.3.
q.e.d.

Now we are in position to characterize the dimension of a non-empty topological manifold M in terms of its local homology. Namely by 6.3 we know that $\dim M = m$ if and only if $\widetilde{SH}_m(M \cup C(M-x);\mathbb{Z}/2) \neq 0$, where x is an arbitrary point in M. If $f : M \to N$ is a homeomorphism, then f can be extended to a homeomorphism $g : M \cup C(M-x) \to N \cup C(N-g(x))$ and so the corresponding local homology groups are isomorphic. Thus
$$\dim M = \dim N.$$

We summarize our discussion with:

Theorem 6.4. *Let $f : M \to N$ be a homeomorphism between non-empty manifolds. Then*
$$\dim M = \dim N.$$

Remark: *Let $Y \subset X$ be a subspace, then the reduced homology of $X \cup C(Y)$ is called the **relative homology** of $Y \subset X$ and it is denoted by*
$$SH_k(X,Y;\mathbb{Z}/2) := \widetilde{SH}_k(X \cup C(Y);\mathbb{Z}/2).$$

4. Exercises

(1) Give a formula for the map $g : D^n \to S^{n-1}$ described in Theorem 6.1 and prove that it is continuous.

(2) Let $A \in M_n(\mathbb{R})$ be a matrix whose entries are positive (non-negative). Show that A has a positive (non-negative) eigenvalue.

Chapter 7

Homology of some important spaces and the Euler characteristic

1. The fundamental class

Given a space X it is very useful to have some explicit non-trivial homology classes. The most important example is the fundamental class of a compact m-dimensional $\mathbb{Z}/2$-oriented regular stratifold $\mathbf{S}$ which we introduced as $[\mathbf{S}]_{\mathbb{Z}/2} := [\mathbf{S}, \mathrm{id}] \in SH_m(\mathbf{S}; \mathbb{Z}/2)$. We have shown that for a sphere the fundamental class is non-trivial. In the following result, we generalize this.

Proposition 7.1. *Let* $\mathbf{S}$ *be a compact* m-*dimensional* $\mathbb{Z}/2$-*oriented regular stratifold with* $\mathbf{S}^m \neq \varnothing$. *Then the fundamental class* $[\mathbf{S}]_{\mathbb{Z}/2} \in SH_m(\mathbf{S}; \mathbb{Z}/2)$ *is non-trivial.*

Proof: The 0-dimensional case is clear and so we assume that $m > 0$. We reduce the proof of the statement to the special case of spheres where it is already known. For this we consider a smooth embedding $\psi : B^m \hookrightarrow \mathbf{S}^m$, where B^m is the open unit ball, and we decompose $\mathbf{S}$ as $\psi(B^m) =: U$ and $\mathbf{S} - \psi(0) =: V$. Then $U \cap V = \psi(B^m - 0)$. We want to determine $d([\mathbf{S}]_{\mathbb{Z}/2})$, where d is the boundary operator in the Mayer-Vietoris sequence corresponding to the covering of $\mathbf{S}$ by U and V. We choose a smooth function $\eta : [0, 1] \to [0, 1]$, which is 0 near 0, 1 near 1 and $\eta(t) = t$ near $1/2$, and then define $\rho : \mathbf{S} \to [0, 1]$ by mapping $\psi(x)$ to $\eta(\|x\|)$ and $\mathbf{S} - \mathrm{im}\,\psi$ to 1. Then

$1/2$ is a regular value of ρ and by definition of the boundary operator we have
$$d([\mathbf{S}]_{\mathbb{Z}/2}) = [\rho^{-1}(1/2), i] \in SH_{m-1}(U \cap V),$$
where $i : \rho^{-1}(1/2) \to U \cap V$ is the inclusion. Thus it suffices to show that $[\rho^{-1}(1/2), i] \neq 0$. Since $\psi|_{\frac{1}{2}S^{m-1}}$ is a homeomorphism from $\frac{1}{2}S^{m-1} = \{x \in \mathbb{R}^m \mid ||x|| = 1/2\}$ to $\psi(\frac{1}{2}S^{m-1}) = \rho^{-1}(1/2)$, we have
$$[\rho^{-1}(1/2), i] = \psi_*[\frac{1}{2}S^{m-1}, \mathrm{id}] = \psi_*[\frac{1}{2}S^{m-1}]_{\mathbb{Z}/2}.$$
The inclusion $\rho^{-1}(1/2) \to U \cap V$ is a homotopy equivalence and thus we are finished since $[\frac{1}{2}S^{m-1}]_{\mathbb{Z}/2} \neq 0$.
q.e.d.

2. $\mathbb{Z}/2$-homology of projective spaces

The most important geometric spaces are the classical Euclidean spaces $\mathbb{R}^n$ and $\mathbb{C}^n$, the home of **affine geometry**. It was an important breakthrough in the history of mathematics when **projective geometry** was invented. The basic idea is to add certain points at infinity to $\mathbb{R}^n$ and $\mathbb{C}^n$. The effect of this change is not so easy to describe. One important difference is that projective spaces are compact. Another is that the intersection of two hyperplanes (projective subspaces of codimension 1) is always non-empty. Many interesting spaces, in particular projective algebraic varieties, are contained in projective spaces so that they are the "home" of algebraic geometry. In topology they play an important role for classifying line bundles and so are at the heart of the theory of characteristic classes.

Many important questions can be formulated and solved using the homology (and cohomology) of projective spaces. Before we compute the homology groups, we have to define projective spaces. They are the set of all lines through 0 in $\mathbb{R}^{n+1}$ or $\mathbb{C}^{n+1}$. The lines which are not contained in $\mathbb{R}^n \times 0$ or $\mathbb{C}^n \times 0$ are in a $1-1$ correspondence with $\mathbb{R}^n$ or $\mathbb{C}^n$, where the bijection maps a point x in $\mathbb{R}^n$ or $\mathbb{C}^n$ to the line given by $(x, 1)$. Thus $\mathbb{R}^n$ resp. $\mathbb{C}^n$ is contained in $\mathbb{RP}^n$ resp. $\mathbb{CP}^n$. The lines which are contained in $\mathbb{R}^n \times 0$ or $\mathbb{C}^n \times 0$ are called points at infinity. They are parametrized by $\mathbb{RP}^{n-1}$ resp. $\mathbb{CP}^{n-1}$. Thus we obtain a decomposition of $\mathbb{RP}^n$ as $\mathbb{R}^n \cup \mathbb{RP}^{n-1}$ and $\mathbb{CP}^n$ as $\mathbb{C}^n \cup \mathbb{CP}^{n-1}$.

To see that the projective spaces are compact, we give a slightly different definition by representing a line by a vector of norm 1.

2. $\mathbb{Z}/2$-homology of projective spaces

We begin with the **complex projective space** $\mathbb{CP}^m$. This may be defined as a quotient space of $S^{2m+1} = \{x = (x_0, \ldots, x_m) \in \mathbb{C}^{m+1} \mid ||x|| = 1\}$, where $||\ ||$ is the standard norm on $\mathbb{C}^{m+1}$, by the equivalence relation $\sim$ where $x \sim y$ if and only if there is a complex number λ such that $\lambda x = y$. In other words two points in S^{2m+1} are equivalent if they span the same line in $\mathbb{C}^{m+1}$. The space $\mathbb{CP}^m$ is a topological manifold of dimension $2m$ and one can introduce a smooth structure in a natural way [**Hi**, p. 14]. Actually, here the coordinate changes are not only smooth maps but holomorphic maps, and thus $\mathbb{CP}^m$ is what one calls a complex manifold, but we don't need this structure and consider $\mathbb{CP}^m$ as a smooth manifold.

To compute its homology, we decompose it into open subspaces
$$U := \{[x] \in \mathbb{CP}^m \mid x_m \neq 0\}$$
and
$$V := \{[x] \in \mathbb{CP}^m \mid |x_m| < 1\}.$$
The reader should check the following properties: U is homotopy equivalent to a point (a homotopy between the identity on U and a constant map is given by $h([x], t) := [tx_0, \ldots, tx_{m-1}, x_m]$), and the inclusion of $\mathbb{CP}^{m-1}$ into V is a homotopy equivalence. A homotopy between the identity on V and a map from V to $\mathbb{CP}^{m-1}$ is given by $h([x], t) := [x_0, \ldots, x_{m-1}, tx_m]$. Furthermore, the intersection $U \cap V$ is homotopy equivalent to S^{2m-1}. The reason is that we actually have a homeomorphism from U to the open unit ball by mapping $[(x_0, \ldots, x_m)]$ to $(x_0/x_m, \ldots, x_{m-1}/x_m)$ and under this homeomorphism $U \cap V$ is mapped to the complement of 0, which is homotopy equivalent to S^{2m-1}.

Thus the homotopy axiom together with the Mayer-Vietoris sequence for $\mathbb{Z}/2$-homology gives an exact sequence:
$$\cdots \to \widetilde{SH}_k(S^{2m-1}; \mathbb{Z}/2) \to \widetilde{SH}_k(\mathbb{CP}^{m-1}; \mathbb{Z}/2)$$
$$\to \widetilde{SH}_k(\mathbb{CP}^m; \mathbb{Z}/2) \to \widetilde{SH}_{k-1}(S^{2m-1}; \mathbb{Z}/2) \to \cdots.$$
Since $\widetilde{SH}_r(S^{2m-1}; \mathbb{Z}/2) = 0$ for $r \neq 2m-1$, we conclude inductively:

Theorem 7.2. $SH_k(\mathbb{CP}^m; \mathbb{Z}/2) \cong \mathbb{Z}/2$ for k even and $k \leq 2m$, and is 0 otherwise. The nontrivial homology class in $SH_{2n}(\mathbb{CP}^m; \mathbb{Z}/2)$ for $n \leq m$ is given by $[\mathbb{CP}^n, i]$, where i is the inclusion from $\mathbb{CP}^n$ to $\mathbb{CP}^m$.

The last statement follows from Proposition 7.1.

To compute the homology of the **real projective space**
$$\mathbb{RP}^m := S^m / x \sim -x,$$

which is a closed smooth m-dimensional manifold [**Hi**, p. 13], we use the same approach as for the complex projective spaces. We decompose $\mathbb{RP}^m$ as $U := \{[x] \in \mathbb{RP}^m \mid x_{m+1} \neq 0\}$ and $V := \{[x] \in \mathbb{RP}^m \mid |x_{m+1}| < 1\}$. A similar argument as above shows: U is homotopy equivalent to a point, and the inclusion from $\mathbb{RP}^{m-1}$ to V is a homotopy equivalence. Furthermore, the intersection $U \cap V$ is homotopy equivalent to S^{m-1}.

The decomposition $\mathbb{RP}^m = U \cup V$ gives an exact sequence:
$$\widetilde{SH}_k(S^{m-1}; \mathbb{Z}/2) \to \widetilde{SH}_k(\mathbb{RP}^{m-1}; \mathbb{Z}/2) \xrightarrow{i_*} \widetilde{SH}_k(\mathbb{RP}^m; \mathbb{Z}/2)$$
$$\to \widetilde{SH}_{k-1}(S^{m-1}; \mathbb{Z}/2) \to \widetilde{SH}_{k-1}(\mathbb{RP}^{m-1}; \mathbb{Z}/2).$$
This implies that for k different from m or $m-1$ the inclusion $\mathbb{RP}^{m-1} \to \mathbb{RP}^m$ induces an isomorphism $i_* : \widetilde{SH}_k(\mathbb{RP}^{m-1}; \mathbb{Z}/2) \to \widetilde{SH}_k(\mathbb{RP}^m; \mathbb{Z}/2)$. Since by Proposition 7.1 $\widetilde{SH}_m(\mathbb{RP}^m; \mathbb{Z}/2) \neq 0$ and $\widetilde{SH}_{m-1}(S^{m-1}; \mathbb{Z}/2) \cong \mathbb{Z}/2$ and by induction $\widetilde{SH}_m(\mathbb{RP}^{m-1}; \mathbb{Z}/2) = 0$, we conclude that $\widetilde{SH}_m(\mathbb{RP}^m; \mathbb{Z}/2) \cong \mathbb{Z}/2$ and that $\widetilde{SH}_m(\mathbb{RP}^m; \mathbb{Z}/2) \to \widetilde{SH}_{m-1}(S^{m-1}; \mathbb{Z}/2)$ is an isomorphism. Thus $i_* : \widetilde{SH}_{m-1}(\mathbb{RP}^{m-1}; \mathbb{Z}/2) \to \widetilde{SH}_{m-1}(\mathbb{RP}^m; \mathbb{Z}/2)$ is injective. Inductively we have shown:

Theorem 7.3. $SH_k(\mathbb{RP}^m; \mathbb{Z}/2) \cong \mathbb{Z}/2$ for $k \leq m$, and 0 otherwise. The nontrivial element in $SH_k(\mathbb{RP}^m; \mathbb{Z}/2) \cong \mathbb{Z}/2$ for $k \leq m$ is given by $[\mathbb{RP}^k, i]_{\mathbb{Z}/2}$ where i is the inclusion from $\mathbb{RP}^k$ to $\mathbb{RP}^m$.

3. Betti numbers and the Euler characteristic

The Betti numbers are important invariants of topological spaces and for some topological spaces X one can use them to define the Euler characteristic.

Definition: *Let X be a topological space. The k-th $\mathbb{Z}/2$-**Betti number** is $b_k(X; \mathbb{Z}/2) := \dim_{\mathbb{Z}/2} SH_k(X; \mathbb{Z}/2)$.*

*A topological space X is called $\mathbb{Z}/2$-**homologically finite**, if for all but finitely many k, the homology groups $SH_k(X; \mathbb{Z}/2)$ are zero, and finite dimensional in the remaining cases.*

*For a $\mathbb{Z}/2$-homologically finite space X, we define the **Euler characteristic** as $e(X) := \sum_i (-1)^i b_i(X; \mathbb{Z}/2)$.*

At the end of this chapter we will prove that all compact smooth manifolds are $\mathbb{Z}/2$-homologically finite and thus their Euler characteristic can be

3. Betti numbers and the Euler characteristic

defined.

The computations in the previous section imply:

i) Suppose $m > 0$. Then $b_k(S^m; \mathbb{Z}/2) = 1$ for $k = 0$ or $k = m$ and 0 otherwise. Thus $e(S^m) = 2$ for m even and $e(S^m) = 0$ for m odd.

ii) $b_k(\mathbb{CP}^m; \mathbb{Z}/2) = 1$ for k even and $0 \leq k \leq 2m$ and $b_k(\mathbb{CP}^m; \mathbb{Z}/2) = 0$ else. Thus $e(\mathbb{CP}^m) = m + 1$.

iii) $b_k(\mathbb{RP}^m; \mathbb{Z}/2) = 1$ for $0 \leq k \leq m$ and $b_k(\mathbb{RP}^m; \mathbb{Z}/2) = 0$ otherwise. Thus $e(\mathbb{RP}^m) = 1$ for m even and $e(\mathbb{RP}^m) = 0$ for m odd.

The significance of the Euler characteristic cannot immediately be seen from the above definition. To indicate its importance we list the following fundamental properties without proof.

i) The Euler characteristic is an obstruction to the existence of nowhere vanishing vector fields on a closed smooth manifold. However, if such a vector field exists, then the Euler characteristic vanishes. We will show that S^m has a nowhere vanishing vector field if and only if m is odd.

ii) The Euler characteristic has to be even if a closed smooth manifold is the boundary of a compact smooth manifold. An example of a closed smooth manifold with odd Euler characteristic is given by one of the examples above, the Euler characteristic of $\mathbb{RP}^{2k}$ is 1. Thus $\mathbb{RP}^{2k}$ is not the boundary of a compact smooth manifold.

iii) For a finite polyhedron, the Euler characteristic can be computed from its combinatorial data: It is the alternating sum of the number of k-dimensional faces.

The following property is very useful for computing the Euler characteristic without knowing the homology.

Theorem 7.4. *Let U and V be $\mathbb{Z}/2$-homologically finite open subspaces of a topological space X, and suppose also that $U \cap V$ is $\mathbb{Z}/2$-homologically finite. Then $U \cup V$ is $\mathbb{Z}/2$-homologically finite and*

$$e(U \cup V) = e(U) + e(V) - e(U \cap V).$$

Proof: The result follows from the Mayer-Vietoris sequence. On the one hand, exactness of the sequence implies that $U \cup V$ is $\mathbb{Z}/2$-homologically finite. The formula is a consequence of the fact we explained earlier: let $0 \to A_n \xrightarrow{f_n} A_{n-1} \xrightarrow{f_{n-1}} \cdots \xrightarrow{f_1} A_0 \to 0$ be an exact sequence of finite dimensional K-vector spaces, where K is some field. Then

$$\sum_{i=0}^{n}(-1)^i \dim A_i = 0.$$

Applying this formula to the exact Mayer-Vietoris sequence, we obtain

$$e(U \cup V) = e(U) + e(V) - e(U \cap V).$$

q.e.d.

We finish this chapter by proving the previously claimed result that compact manifolds are $\mathbb{Z}/2$-homologically finite.

Theorem 7.5. *A compact smooth c-manifold is $\mathbb{Z}/2$-homologically finite.*

Proof: It is enough to prove this for closed manifolds since the case of non-empty boundary can be reduced to the closed case as we now explain. Let W be a compact c-manifold with non-empty boundary. The double $W \cup_{\partial W} W$ is a closed manifold. Assuming the closed case, $W \cup_{\partial W} W$ and ∂W are $\mathbb{Z}/2$-homologically finite. We decompose $W \cup_{\partial W} W$ as $U \cup V$, where U is the union of one copy of W together with the bicollar used in the gluing and V is the union of the other copy of W together with the bicollar. The spaces U and V are both homotopy equivalent to W and $U \cap V$ is homotopy equivalent to ∂W and so a similar argument as in the proof of Theorem 7.4 shows that if $U \cup V$ and $U \cap V$ are $\mathbb{Z}/2$-homologically finite, then U and V are $\mathbb{Z}/2$-homologically finite. Since W is homotopy equivalent to U (or V), it has the same homology groups and so is $\mathbb{Z}/2$-homologically finite.

To prove the theorem for a closed manifold M, we embed M into $\mathbb{R}^N$ for some N [**Hi**, Theorem I.3.4] and consider a tubular neighbourhood U ([**Hi**] Theorem IV.5.2). Let $r : U \to M$ be the projection of the normal bundle to M which is a retraction onto M, that is, $r(x) = x$ for all $x \in M$. Now we choose for each point $x \in M$ an open cube in U containing x. Since M is compact, we can cover M by finitely many cubes C_i:

$$M \subset \bigcup C_i \subset U.$$

The union of finitely many open cubes is $\mathbb{Z}/2$-homologically finite. This follows inductively. It is clear for a single cube. We suppose that the union of $k-1$ cubes is $\mathbb{Z}/2$-homologically finite. If we add another cube then the intersection of the new cube with the union of the $k-1$ cubes is a union of at most $k-1$ cubes since the intersection of two cubes is again a cube or empty. Thus Theorem 7.4 implies that the union of k cubes is $\mathbb{Z}/2$-homologically finite.

Since $r|_{\bigcup C_i}$ is a retraction, we conclude that the homology groups of $\bigcup C_i$ are mapped surjectively onto the homology groups of M, which finishes the argument.
q.e.d.

4. Exercises

(1) Compute the Euler characteristic of the 2-torus, or more generally of any space of the form $X \times S^1$.

(2) Let X be a topological space and let $\{U_1, U_2, \ldots, U_n\}$ be an open covering such that all intersections are $\mathbb{Z}/2$-homologically finite spaces. Show that:
$$e(X) = \sum_i e(U_i) - \sum_{i<j} e(U_i \cap U_j)$$
$$+ \sum_{i<j<k} (e(U_i \cap U_j \cap U_k) + \cdots + (-1)^{n+1} e(U_1 \cap U_2 \cdots \cap U_n)).$$

(3) Let M be a compact smooth manifold, and $\widetilde{M} \to M$ be a finite covering space. Show that if the preimage of each point consists of k points than $e(\widetilde{M}) = k \cdot e(M)$. Deduce that if G is a finite group of order k and k doesn't divide $e(N)$ then there is no free G-action on the compact smooth manifold N. Show that the only group acting freely on S^{2n} is $\mathbb{Z}/2$. Show that if N has a free S^1-action then $e(N) = 0$. Can you classify all groups with a free action on the real projective plane?

Chapter 8

Integral homology and the mapping degree

Prerequisites: The only new ingredient used in this chapter is the definition of the orientation of smooth manifolds, which can be found in [**B-J**] or [**Hi**].

1. Integral homology groups

In this chapter, we will introduce integral stratifold homology. This is a most powerful tool in topology which is fundamental for studying all sorts of problems. The definition is completely analogous to that of $\mathbb{Z}/2$-homology, the only difference being that we require the top-dimensional strata of our stratifolds to be oriented.

Definition: *An* **oriented** *m-dimensional c-stratifold is an m-dimensional c-stratifold $\mathbf{T}$ with $\overset{\circ}{\mathbf{T}}{}^{m-1} = \varnothing$ and an orientation on $\overset{\circ}{\mathbf{T}}{}^{m}$.*

An orientation on $\mathbf{T}$ induces an orientation of $\partial \mathbf{T}$ which is fixed by requiring that the collar of $\mathbf{T}$ preserves the product orientation on $(\partial \mathbf{T})^{m-1} \times (0, \epsilon)$. If we reverse the orientation of $\overset{\circ}{\mathbf{T}}{}^{m}$, we call the corresponding oriented stratifold $-\mathbf{T}$.

We would like to note that there are different ways to orient the boundary of a smooth manifold W in the literature. Our convention is equivalent to the one which characterizes the orientation of the boundary at a point $x \in \partial W$ in the boundary by requiring that a basis of the tangent space of the boundary at x is compatible with the orientation we want to define, if this basis followed by an inward pointing normal vector is the given orientation of W at x. An other often used convention is that an outward pointing normal vector followed by the oriented basis we want to define is the orientation of W at x. This convention differs from ours by the sign $(-1)^n$, where $n = \dim W$. The orientation convention plays later on a role, when we define the boundary operator in the Mayer-Vietoris sequence, where one would obtain a different operator from ours differing by a sign $(-1)^n$, where n is the degree of the homology group on which the operator is defined.

In complete analogy with the case of smooth manifolds, we define bordism groups of compact oriented m-dimensional regular stratifolds denoted $SH_m(X)$:

$$SH_m(X) := \{(\mathbf{S}, g)\}/\text{bord},$$

where $\mathbf{S}$ is an m-dimensional compact, oriented, regular stratifold and $g : \mathbf{S} \to X$ is a continuous map. The relation "bord" means that two such pairs $(\mathbf{S}, g)$ and $(\mathbf{S}', g')$ are equivalent if there is a compact oriented regular c-stratifold $\mathbf{T}$ with boundary $\mathbf{S} \sqcup (-\mathbf{S}')$ and $g \sqcup g'$ extends to a map $G : \mathbf{T} \to X$. The role of the negative orientation on $\mathbf{S}'$ is the following. To show that the relation is transitive, we proceed as for $\mathbb{Z}/2$-homology and glue a bordism $\mathbf{T}$ between $\mathbf{S}$ and $\mathbf{S}'$ and a bordism $\mathbf{T}'$ between $\mathbf{S}'$ and $\mathbf{S}''$ along $\mathbf{S}'$. We have to guarantee that the orientations on the top stratum of $\mathbf{T}$ and of $\mathbf{T}'$ fit together to give an orientation of the top stratum of $\mathbf{T} \cup_{\mathbf{S}'} \mathbf{T}'$. This is the case if the orientations on $\mathbf{S}'$ induced from $\mathbf{T}$ and $\mathbf{T}'$ are opposite.

With this clarification the proof that the relation is an equivalence relation is the same as for $\mathbb{Z}/2$-homology (Proposition 4.4). It is useful to note that $-[\mathbf{S}, f] = [-\mathbf{S}, f]$, i.e., the inverse of $(\mathbf{S}, f)$ is given by changing the orientation of $\mathbf{S}$.

If $f : X \to Y$ is continuous, we define $f_* : SH_m(X) \to SH_m(Y)$ by composition just as we did for $\mathbb{Z}/2$-homology. In this way we obtain functors from spaces to abelian groups and these functors again form a homology theory. This means that homotopic maps induce the same map in integral homology and that there is a Mayer-Vietoris sequence commuting with induced maps (for the definition of a homology theory see also the next chapter). The construction of the boundary operator in the Mayer-Vietoris sequence which we gave for $\mathbb{Z}/2$-homology extends once we convince ourselves that the constructions used there (like cutting and gluing) transform oriented regular stratifolds into oriented regular stratifolds. But these facts are obvious once we have fixed an orientation on the preimage of a regular value s of a smooth map $f : M \to \mathbb{R}$ on an oriented manifold M. We orient such a preimage by requiring that the orientation of it together with a vector v in the normal bundle to $f^{-1}(s)$ is an orientation of M, if the image of v under the differential of f is positive. Further we note that as for $\mathbb{Z}/2$-homology one can define **reduced stratifold homology groups** $\widetilde{SH}_k(X)$ as the kernel of the map from $SH_k(X)$ to $SH_k(pt)$ and that one has an analogous Mayer-Vietoris sequence for reduced homology groups.

Theorem 8.1. *The functor which assigns the abelian group $SH_m(X)$ to the space X defines a homology theory. This functor is called* **integral stratifold homology** *or, for short,* **integral homology**.

To determine the integral homology groups of a point, we first note that for $m > 0$ the cone over an oriented regular stratifold **S** is an oriented regular stratifold with boundary **S**. Thus for $m > 0$ we have $SH_m(\text{pt}) = 0$. To determine $SH_0(\text{pt})$, we remind the reader that an orientation of a 0-dimensional manifold assigns to each point x a number $\epsilon(x) \in \pm 1$, and that the boundary of an oriented interval $[a,b]$ has an induced orientation such that $\epsilon(a) = -\epsilon(b)$ [**B-J**]. Thus, if a compact 0-dimensional manifold M is the boundary of a compact oriented 1-dimensional manifold, then $\sum_{x \in M} \epsilon(x) = 0$. In turn, if $\sum_{x \in M} \epsilon(x) = 0$, then we can group the points of M into pairs with opposite orientation and take as null bordism for these pairs a union of intervals. Since oriented regular stratifolds of dimension 0 and 1 are the same as oriented manifolds, we conclude:

Theorem 8.2. *The map*
$$SH_0(\text{pt}) \to \mathbb{Z}$$
mapping $[M,g]$ to $\sum_{x \in M} \epsilon(x)$ is an isomorphism. Furthermore for $m \neq 0$ we have
$$SH_m(\text{pt}) = 0.$$

Since an oriented regular stratifold is automatically $\mathbb{Z}/2$-oriented we have a forgetful homomorphism
$$SH_k(X) \to SH_k(X; \mathbb{Z}/2).$$
We will discuss this homomorphism at the end of this chapter.

As with $\mathbb{Z}/2$-homology, we say that a space X is **homologically finite** if for all but finitely many k, the homology groups $SH_k(X)$ are zero and the remaining homology groups are finitely generated. The same argument as for $\mathbb{Z}/2$-homology implies that compact smooth manifolds are homologically finite. We define the **Betti numbers** $b_k(X)$ as the rank of $SH_k(X)$. This is an important invariant of spaces. We recall from algebra that the rank of an abelian group G is equal to the dimension of the $\mathbb{Q}$-vector space $G \otimes \mathbb{Q}$ (for some basic information about tensor products, see Appendix C). It is useful here to remind the reader of the **fundamental theorem for finitely generated abelian groups** G, which says that G is isomorphic to $\mathbb{Z}^r \oplus \text{tor}(G)$, where $\text{tor}(G) = \{g \in G \mid ng = 0 \text{ for some natural number } n \neq 0\}$ is the torsion subgroup of G. Since $\text{tor}(G) \otimes \mathbb{Q} = 0$, the number r is equal to the rank of G. The torsion subgroup T is itself isomorphic to a sum of finite cyclic groups: $\text{tor}(G) \cong \bigoplus_i \mathbb{Z}/n_i$. If X is homologically finite, then $b_k(X)$ is zero for all but finitely many k and finite otherwise.

Using the Mayer-Vietoris sequence, one computes the integral homology of the sphere S^m for $m > 0$ as for $\mathbb{Z}/2$-homology. The result is:

$$SH_k(S^m) \cong \begin{cases} \mathbb{Z} & k = 0, m \\ 0 & \text{otherwise.} \end{cases}$$

A generator of $SH_m(S^m)$ is given by the homology class $[S^m, \text{id}]$. Here we orient S^m as the boundary of D^{m+1}, which we equip with the orientation induced from the standard orientation of $\mathbb{R}^{m+1}$. (Note that this orientation on S^m is characterized by the property that a basis of $T_x S^m$ belongs to the orientation if and only if it gives the standard orientation of $\mathbb{R}^{m+1}$ when followed by an inward pointing normal vector.)

As a first important application of integral homology we define the degree of a map from a compact, oriented, m-dimensional, regular stratifold to a connected, oriented, m-dimensional, smooth manifold M. We start with the definition of the fundamental class.

Definition: *Let $\mathbf{S}$ be a compact oriented m-dimensional regular stratifold. The **fundamental class** of $\mathbf{S}$ is $[\mathbf{S}, \text{id}] \in SH_m(\mathbf{S})$. We abbreviate it as $[\mathbf{S}] := [\mathbf{S}, \text{id}]$.*

If we change the orientation of $\mathbf{S}$ passing to $-\mathbf{S}$, then the fundamental class changes orientation as well: $[-\mathbf{S}] = -[\mathbf{S}]$. Under the homomorphism $SH_m(\mathbf{S}) \to SH_m(\mathbf{S}; \mathbb{Z}/2)$, the fundamental class maps to the $\mathbb{Z}/2$-fundamental class: $[\mathbf{S}] \mapsto [\mathbf{S}]_{\mathbb{Z}/2}$. This implies that the fundamental class is non-trivial. But one actually knows more:

Theorem 8.3. *Let* **S** *be a compact oriented m-dimensional regular stratifold. Then $k[\mathbf{S}] \in SH_m(\mathbf{S})$ is non-trivial for all $k \in \mathbb{Z} - \{0\}$ (we say that $[\mathbf{S}]$ has infinite order) and $[\mathbf{S}]$ is primitive (i.e., not divisible by any $r > 1$).*

Proof: The proof is similar to the proof of Proposition 7.1. The case $m = 0$ is trivial. For $m > 0$, we take an orientation-preserving embedding $i : D^m \to \mathbf{S}^m$ of D^m into the top stratum of **S**. As in the proof of Proposition 7.1 this gives rise to a decomposition $\mathbf{S} = U \cup V$ with $U = i(\overset{\circ}{D^m})$ and $V = \mathbf{S} - i(0)$. The associated Mayer-Vietoris sequence gives a homomorphism from $SH_m(\mathbf{S}) = \widetilde{SH}_m(\mathbf{S}) \to \widetilde{SH}_{m-1}(S^{m-1})$ mapping $[\mathbf{S}]$ to $[S^{m-1}]$ (where we have oriented S^{m-1} as the boundary of D^m). The statement now follows by induction.
q.e.d.

2. The degree

Now we define the **degree** and begin by defining it only for maps from compact oriented m-dimensional regular stratifolds **S** to S^m. Recall that we have $SH_m(S^m) \cong \mathbb{Z}$ generated by $[S^m]$ for all $m > 0$.

Definition: *Let* **S** *be a compact oriented m-dimensional regular stratifold, $m > 0$, and $f : \mathbf{S} \to S^m$ be a continuous map. Then we define*
$$\deg f := k \in \mathbb{Z}$$
where $[\mathbf{S}, f] = k[S^m]$.

In other words, $f_*([\mathbf{S}]) = \deg(f)[S^m]$. By construction, homotopic maps have the same degree. For $h : S^m \to S^m$, we see that $h_* : SH_m(S^m) \to SH_m(S^m)$ is multiplication by $\deg h$. As a consequence, we conclude that the degree of the composition of two maps $f, g : S^n \to S^n$ is the product of the degrees:
$$\deg(fg) = \deg(f)\deg(g).$$

One can generalize the definition of the degree to maps from **S** to a connected oriented m-dimensional smooth manifold M: namely one chooses an orientation-preserving embedding of a disc D^m into M and considers the map $p : M \to S^m = D^m/S^{m-1}$ which is the identity on $\overset{\circ}{D^m}$ and maps the rest of M to the point S^{m-1}/S^{m-1}. Then we define the degree of $f : \mathbf{S} \to M$ as
$$\deg(f) := \deg pf.$$
Since any two orientation-preserving embeddings of D^m into M are isotopic [**B-J**], the definition of the degree of f is independent of the choice of this

embedding.

To get a feeling for the degree, we compute it for the map $z^m : S^1 \to S^1$, where we consider S^1 as a subspace of $\mathbb{C}$ and map z to z^m. The degree of z^m is k, where $[S^1, z^m] = k \cdot [S^1] \in SH_1(S^1)$. We will show that $k = m$. We have to construct a bordism between $[S^1, z^m]$ and $m \cdot [S^1]$. The following figure and commentary which follows explain how this can be done.

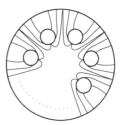

Here we remove $|m|$ small open balls of equal radius, $B_1, \ldots B_m$, from D^2. The centers of the balls are equally distributed around a circle concentric with the boundary of D^2. The space obtained from D^2 by removing these balls is a bordism between $\partial D^2 \cong S^1$ and $\bigcup_{i=1}^m \partial B_m \cong \underbrace{S^1 \sqcup \cdots \sqcup S^1}_{|m|}$. To construct a map from this bordism to S^1, we map the curved lines joining the small circles with the large circle to the image of the endpoint in the large circle under the map z^m. We extend this to a map on the whole bordism by mapping the rest constantly to $1 \in S^1$. If m is positive, this induces the identity map id $: S^1 \to S^1$ on each small circle. Thus we conclude $[S^1, z^m] = m \cdot [S^1, \mathrm{id}] = m \cdot [S^1]$. If m is negative, the induced map on each circle is $z^{-1} = \bar{z} = (z_1, -z_2)$. Thus for $m < 0$ we obtain that the degree of z^m is $-m \cdot \deg z^{-1}$. The degree of z^{-1} is -1. To see this we prove that $[S^1, z] = -[S^1, z^{-1}]$. A bordism between these two objects is given by $W := (S^1 \times [0, 1/2]) \cup_{z^{-1}} ((-S^1) \times [1/2, 1])$ and the map which is given by $z \cup z^{-1}$. The point here is that z^{-1} reverses the orientation (why?) and thus is an orientation-preserving diffeomorphism between S^1 and $(-S^1)$ giving $\partial W = S^1 \sqcup S^1$, where both S^1's have the same orientation. We summarize with

Proposition 8.4. *The degree of $z^m : S^1 \to S^1$ is m.*

From this one can deduce the **fundamental theorem of algebra**.

Theorem 8.5. *Each complex polynomial $f : \mathbb{C} \to \mathbb{C}$ of positive degree has a zero.*

Proof: We can assume that $f(z) = a_0 + a_1 z + \cdots + a_{n-1} z^{n-1} + z^n$. If $a_0 = 0$, then $z = 0$ is a zero, and so we assume $a_0 \neq 0$. We assume that

f has no zero and consider the map $S^1 \to S^1$, $z \mapsto f(z)/|f(z)|$. This map is homotopic to $a_0/|a_0|$ under the homotopy $f(tz)/|f(tz)|$. On the other hand, it is also homotopic to z^n under the following homotopy. For $t \neq 0$ we take $f(t^{-1}z)/|f(t^{-1}z)|$. As t tends to 0, this map tends to z^n. We obtain a contradiction since the degree of $a_0/|a_0|$ is zero while the degree of z^n is n by Proposition 8.4.
q.e.d.

Consider now the reflection on $S^1 \subset \mathbb{C}$, which maps a complex number $z = (z_1, z_2)$ to $(z_1, -z_2) = \bar{z}$. Since $\bar{z} = z^{-1}$, we conclude from Proposition 8.4 that the degree of this reflection is -1. Using the inductive computation of $SH_m(S^m)$, we conclude that the degree of the reflection map $S^m \to S^m$ mapping $(z_1, z_2, \ldots, z_{m+1}) \longmapsto (z_1, -z_2, z_3, \ldots, z_{m+1})$ is also -1. Since all reflection maps

$$s_i : S^m \to S^m$$

mapping $(z_1, \ldots, z_{m+1})$ to $(z_1, \ldots, z_{i-1}, -z_i, z_{i+1}, \ldots, z_{m+1})$ are conjugate to s_2, we conclude that for each i the degree of s_i is -1. Since $-\mathrm{id} = s_1 \circ \cdots \circ s_{m+1}$, we conclude

Proposition 8.6. *For $m > 0$ the degree of the antipodal map on S^m is $(-1)^{m+1}$.*

As a consequence, for m even the identity is not homotopic to $-\mathrm{id}$. This fact leads to the answers to an important question: Which spheres admit a nowhere vanishing continuous vector field? Recall that the tangent bundle of S^m is $TS^m = \{(x, w) \in S^m \times \mathbb{R}^{m+1} \mid w \perp x\}$. For those who are not familiar with tangent bundles, we suggest taking the right side as a definition. But we also suggest that you convince yourself that for each x the vectors w with $w \perp x$ fit with the intuitive notion of the tangent space of S^m at x.

A continuous **vector field** on a smooth manifold M is a continuous map $v : M \to TM$ such that $pv = \mathrm{id}$, where p is the projection of the tangent bundle. In the case of the sphere a nowhere vanishing continuous vector field is the same as a map $v : S^m \to \mathbb{R}^{m+1} - \{0\}$ with $v(x) \perp x$ for all $x \in S^m$. Replacing $v(x)$ by $v(x)/\|v(x)\|$, we can assume that $v(x) \in S^m$ for all $x \in S^m$. If S^m admits a nowhere zero continuous vector field then $H : S^m \times I \to S^m$ mapping $(x, t) \mapsto (cos(\pi \cdot t))x + (sin(\pi \cdot t)) \cdot v(x)$ is a homotopy between id and $-\mathrm{id}$. But this is not possible if m is even. Thus we have proven

Theorem 8.7. *There is no nowhere vanishing continuous vector field on S^{2k}.*

For S^2 this result runs under the name of the **hedgehog theorem** and says that it is impossible to comb the spines of a hedgehog continuously.

On S^{2k+1} there is a nowhere vanishing vector field, for example:
$$v(x_1, x_2, \ldots, x_{2k+1}, x_{2k+2}) := (-x_2, x_1, -x_4, x_3, \ldots, -x_{2k+2}, x_{2k+1}),$$
or in complex coordinates
$$v(z_1, \ldots, z_{k+1}) := (iz_1, \ldots, iz_{k+1}).$$

Thus we have shown:

There exists a nowhere vanishing vector field on S^m if and only if m is odd.

Remark: This is a special case of a much more general theorem: there is a nowhere vanishing vector field on a compact m-dimensional smooth manifold M if and only if the Euler characteristic $e(M)$ vanishes. Note that this is consistent with our previous calculation that the Euler characteristic of S^m is 0 if m is odd, and 2 if m is even.

3. Integral homology groups of projective spaces

We want to compute the integral homology of our favorite spaces. We recall that for $m > 0$ we have
$$SH_k(S^m) \cong \begin{cases} \mathbb{Z} & k = 0, m \\ 0 & \text{otherwise.} \end{cases}$$

We treat complex projective spaces inductively as we did for $\mathbb{Z}/2$-homology. Using the decomposition of $\mathbb{CP}^m$ into U and V as in the proof of Theorem 7.2 we conclude from the Mayer-Vietoris sequence:

Theorem 8.8. $SH_k(\mathbb{CP}^m) \cong \mathbb{Z}$ *for k even and $0 \leq k \leq 2m$ and 0 otherwise. The non-trivial homology class in $SH_{2n}(\mathbb{CP}^m)$ for $n \leq m$ is given by $[\mathbb{CP}^n, i]$, where i is the inclusion from $\mathbb{CP}^n$ to $\mathbb{CP}^m$.*

Finally we compute the integral homology of $\mathbb{RP}^m$.

3. Integral homology groups of projective spaces

Theorem 8.9. $SH_k(\mathbb{RP}^m) \cong \mathbb{Z}$ for $k = 0$ and $k = m$, if m is odd. $SH_k(\mathbb{RP}^m) \cong \mathbb{Z}/2$ for k odd and $k < m$. The other homology groups are zero. Generators of the non-trivial homology groups for k odd are represented by $[\mathbb{RP}^k, i]$, where i is the inclusion.

Proof: Again we use from §7 the decomposition of $\mathbb{RP}^m$ into U and V with U homotopy equivalent to a point, V homotopy equivalent to $\mathbb{RP}^{m-1}$, and $U \cap V$ homotopy equivalent to S^{m-1}. Then we conclude from the Mayer-Vietoris sequence by induction that for $k < m - 1$ we have isomorphisms

$$i_* : SH_k(\mathbb{RP}^{m-1}) \cong SH_k(\mathbb{RP}^m).$$

To finish the induction we consider the exact sequence obtained from the Mayer-Vietoris sequence

$$0 \to \widetilde{SH}_m(\mathbb{RP}^m) \to \widetilde{SH}_{m-1}(S^{m-1})$$

$$\to \widetilde{SH}_{m-1}(\mathbb{RP}^{m-1}) \to \widetilde{SH}_{m-1}(\mathbb{RP}^m) \to 0.$$

If m is odd, we conclude by induction that $SH_m(\mathbb{RP}^m) \cong \mathbb{Z}$ and from Theorem 8.3 that $[\mathbb{RP}^m] = [\mathbb{RP}^m, \text{id}]$ is a generator. Here we use the fact that $\mathbb{RP}^m$ is orientable and we orient it in such a way that $dp_x : T_x S^m \to T_x \mathbb{RP}^m$ is orientation-preserving. Since by induction $SH_{m-1}(\mathbb{RP}^{m-1}) = 0$, we have $SH_{m-1}(\mathbb{RP}^m) = 0$.

If m is even, we first note that $2i_*([\mathbb{RP}^{m-1}]) = 0$. The reason is that the reflection $r([x_1, \ldots, x_m]) := [-x_1, x_2, \ldots, x_m]$ is an orientation-reversing diffeomorphism of $\mathbb{RP}^{m-1}$. Thus $[\mathbb{RP}^{m-1}] = [-\mathbb{RP}^{m-1}, r] = -r_*([\mathbb{RP}^{m-1}])$. Now consider the homotopy $h([x], t) := [\cos(\pi t)x_1, x_2, \ldots, x_m, \sin(\pi t)x_1]$ between i and ir. Thus $i_*([\mathbb{RP}^{m-1}]) = -i_*([\mathbb{RP}^{m-1}])$.

Next we note that $i_*([\mathbb{RP}^{m-1}]) \neq 0$ since it represents a non-trivial element in $\mathbb{Z}/2$-homology by Theorem 7.3, i.e., it is not even the boundary of a non-oriented regular stratifold with a map to $\mathbb{RP}^m$. Then the statement follows from the exact Mayer-Vietoris sequence above. We already know that $SH_{m-1}(S^{m-1})$ and $SH_{m-1}(\mathbb{RP}^{m-1})$ are infinite cyclic and that $i_*([\mathbb{RP}^{m-1}]) \in SH_{m-1}(\mathbb{RP}^m)$ is non-trivial. Thus $SH_{m-1}(\mathbb{RP}^m)$ is cyclic of order 2 generated by $i_*([\mathbb{RP}^{m-1}])$ and the map $SH_{m-1}(S^{m-1}) \to SH_{m-1}(\mathbb{RP}^{m-1})$ is non-trivial which implies that $SH_m(\mathbb{RP}^m) = 0$.
q.e.d.

4. A comparison between integral and $\mathbb{Z}/2$-homology

An oriented stratifold **S** is automatically $\mathbb{Z}/2$-oriented. Thus we have a homomorphism
$$r : SH_n(X) \longrightarrow SH_n(X; \mathbb{Z}/2)$$
for each topological space X and each n. One often calls it **reduction mod 2**. This homomorphism commutes with induced maps and the boundary operators, i.e., if $f : X \longrightarrow Y$ is a continuous map, then
$$f_* r = r f_* : SH_n(X) \longrightarrow SH_n(Y; \mathbb{Z}/2),$$
and if $X = U \cup V$, then
$$r d_{\mathbb{Z}} = d_{\mathbb{Z}/2} r : SH_n(U \cup V) \longrightarrow SH_{n-1}(U \cap V; \mathbb{Z}/2),$$
where $d_{\mathbb{Z}/2}$ and $d_{\mathbb{Z}}$ are respectively the boundary operators in the Mayer-Vietoris sequences for $\mathbb{Z}/2$-homology and integral homology. A map r (for each space X and each n) fulfilling these two properties is called a **natural transformation** from the functor integral homology to the functor $\mathbb{Z}/2$-homology. Below and in the next chapter, we will consider other natural transformations.

If we want to use r to compare integral homology with $\mathbb{Z}/2$-homology, we need information about the kernel and cokernel of r. The answer is given in terms of an exact sequence, the **Bockstein sequence**.

Theorem 8.10. *There is a natural transformation*
$$d : SH_n(X; \mathbb{Z}/2) \longrightarrow SH_{n-1}(X)$$
and, if X is a smooth manifold or a finite CW-complex (as defined in the next chapter), then the following sequence is exact:
$$\cdots \to SH_n(X) \xrightarrow{\cdot 2} SH_n(X) \xrightarrow{r} SH_n(X; \mathbb{Z}/2) \xrightarrow{d} SH_{n-1}(X) \longrightarrow \cdots.$$

Since we will not apply the Bockstein sequence in this book, we will not give a proof. At the end of this book, we will explain the relation between our definition of homology and the classical definition using singular chains. The groups are naturally isomorphic if X is a smooth manifold or a finite CW-complex (we will define finite CW-complexes in the next chapter). We will prove this in §20. If one uses the classical approach, the proof of the existence of the Bockstein sequence is simple and it actually is a special case of a more general result. Besides reflecting different geometric aspects, the two definitions of homology groups both have specific strengths and weaknesses. For example, the description of the fundamental class of a closed smooth (oriented) manifold is simpler in our approach whereas the Bockstein sequence is more complicated.

The Bockstein sequence gives an answer to a natural question. Let X be a topological space such that all Betti numbers $b_k(X)$ are finite and only finitely many are non-zero. Then one can consider the alternating sum
$$\sum_k (-1)^k b_k(X).$$
The question is what is the relation between this expression and the Euler characteristic
$$e(X) = \sum_k (-1)^k b_k(X; \mathbb{Z}/2).$$

Theorem 8.11. *Let X be a compact smooth manifold or a finite CW-complex. Then $b_k(X)$ is finite and non-trivial only for finitely many k, and*
$$e(X) = \sum_k (-1)^k b_k(X).$$

Proof: We decompose $SH_k(X) \cong \mathbb{Z}^{r(k)} \oplus \mathbb{Z}/2^{a_1} \oplus \cdots \oplus \mathbb{Z}/2^{a_{s(k)}} \oplus T$, where T consists of odd torsion elements. Then the kernel of multiplication with 2 is $(\mathbb{Z}/2)^{s(k)}$ and the cokernel is
$$(\mathbb{Z}/2)^{s(k)} \oplus (\mathbb{Z}/2)^{r(k)}.$$
Thus we have a short exact sequence
$$0 \to (\mathbb{Z}/2)^{s(k)} \oplus (\mathbb{Z}/2)^{r(k)} \to SH_k(X; \mathbb{Z}/2) \to (\mathbb{Z}/2)^{s(k-1)} \to 0$$
implying that $\dim SH_k(X; \mathbb{Z}/2) = s(k) + s(k-1) + r(k)$, and from this we conclude the theorem by a cancellation argument.
q.e.d.

5. Exercises

(1) Let T and T' be two oriented c-stratifolds with $\partial T = -\partial T'$. Show that there is a unique orientation on $T \cup_{\partial T} T'$ which restricts to the orientations on T and T'.

(2) Answer questions 5 and 6 in chapter 5 but now for integral homology.

(3) Let X be a pointed path connected space. Look at the definition of the fundamental group in any textbook. Show that there is a natural map $\pi_1(X, *) \to SH_1(X)$ and that it is surjective with kernel containing the commutator subgroup of $\pi_1(X, *)$. Show that the kernel actually equals the commutator subgroup.

(4) Let $f : M \to N$ be a smooth map between two closed oriented manifolds. Assume that the preimage of some regular value y consists of n points $\{x_1, x_2, \ldots, x_n\}$. If x is a regular value then the differential map $df|_x$ is an isomorphism of vector spaces. We define $\text{sign}(f|_x)$ to be 1 if this isomorphism is orientation-preserving and -1 otherwise. Show that the degree of f is equal to $\sum \text{sign}(f|_{x_k})$. (Hint: Start by showing this in the case $N = S^k$ and $n = 1$.)

(5) Let M be a closed oriented manifold and let $\pi : \widetilde{M} \to M$ be a covering map with fibre consisting of k points. Show that if π is orientation-preserving then the degree of π is k and if π is orientation-reversing then the degree is $-k$.

(6) Let $f : X \to Y$ be a map between two spaces. Define the **mapping cone** $C_f = CX \cup_f Y$. There is a map $f' : X \to C_f$ induced by the inclusion from X to CX. Show that $f'_* = 0$.

(7) Compute the integral homology of the mapping cone of the map $f : S^n \to S^n$ of degree k.

(8) Let $\mathbf{S}$ be a stratifold and $\mathbf{S}'$ an oriented stratifold which is homeomorphic to $\mathbf{S}$. Does it follow that $\mathbf{S}$ is orientable? Is this the case if the codimension 1-stratum of $\mathbf{S}$ is empty?

(9) The spaces $\mathbb{RP}^2$ and $S^1 \vee S^2$ have the same $\mathbb{Z}/2$ homology. Are they homotopy equivalent? Is there a map between those spaces inducing this isomorphism?

(10) Give an example of a map between two spaces $f : X \to Y$ such that $f_* : \widetilde{SH}_k(X) \to \widetilde{SH}_k(Y)$ is the zero map, but f is not null homotopic.

(11) Show that for every continuous map $f : S^{2n} \to S^{2n}$ there is a point $x \in S^{2n}$ such that $f(x) = x$ or $f(x) = -x$. (Hint: Otherwise construct a homotopy between the identity map and the map $-\text{id} : x \mapsto -x$.) Deduce that every continuous map $f : \mathbb{RP}^{2n} \to \mathbb{RP}^{2n}$ has a fixed point.

(12) Let M be a manifold and $f : M \to M$ a diffeomorphism. Define the mapping torus M_f to be $M_f = M \times I / (x, 0) \sim (f(x), 1)$.
a) Give a smooth structure to M_f.
b) Compute $H_k(M_f)$ using the Mayer-Vietoris sequence in terms of the homology of M and f_*.
c) Show that if M is oriented and f orientation-preserving then the product orientation on $M \times (0, 1)$ can be extended to an orientation on M_f.
d) Suppose that M_f is orientable; is M orientable and f orientation-preserving?

(13) Let $\begin{pmatrix} a & b \\ c & d \end{pmatrix} \in SL(2, \mathbb{Z})$ and $f(A) : S^1 \times S^1 \to S^1 \times S^1$ be defined by $(x, y) \mapsto (x^a y^b, x^c y^d)$. Compute $H_k((S^1 \times S^1)_{f(A)})$.

Chapter 9

A comparison theorem for homology theories and CW-complexes

1. The axioms of a homology theory

We have already constructed two homology theories. We now give a general definition of a homology theory.

Definition: *A* **homology theory** *h assigns to each topological space X a sequence of abelian groups $h_n(X)$ for $n \in \mathbb{Z}$, and to each continuous map $f : X \to Y$ homomorphisms $f_* : h_n(X) \to h_n(Y)$. One requires that the following properties hold:*

i) $\mathrm{id}_* = \mathrm{id}$, $(gf)_* = g_* f_*$, i.e., h is a **functor**,

ii) if f is homotopic to g, then $f_* = g_*$, i.e., h is **homotopy invariant**,

iii) for open subsets U and V of X there is a long exact sequence (**Mayer-Vietoris sequence**)

$$\cdots \to h_n(U \cap V) \to h_n(U) \oplus h_n(V) \to h_n(U \cup V)$$

$$\xrightarrow{d} h_{n-1}(U \cap V) \longrightarrow h_{n-1}(U) \oplus h_{n-1}(V) \to \cdots$$

commuting with induced maps (the Mayer-Vietoris sequence is natural). Here the map $h_n(U \cap V) \to h_n(U) \oplus h_n(V)$ is $\alpha \mapsto ((i_U)_*(\alpha), (i_V)_*(\alpha))$, the map $h_n(U) \oplus h_n(V) \to h_n(U \cup V)$ is $(\alpha, \beta) \mapsto (j_U)_*(\alpha) - (j_V)_*(\beta)$ and the map d is a group homomorphism called the **boundary operator**. Note that d is an essential part of the homology theory.

The maps i_U and i_V are the inclusions from $U \cap V$ to U and V, the maps j_U and j_V are the inclusions from U and V to $U \cup V$. The Mayer-Vietoris sequence extends arbitrarily far to the left and to the right.

As before we say that h_n is a **functor** from the category of topological spaces and continuous maps to the category of abelian groups and group homomorphisms. If one requires that $h_n(X) = 0$ for $n < 0$, such a theory is called a **connective homology theory**. The additional requirement that $h_n(\text{pt}) = 0$ for $n \neq 0$ is called the **dimension axiom**. A homology theory is called an **ordinary homology theory** if it satisfies the dimension axiom and a **generalised homology theory** if it does not.

2. Comparison of homology theories

We want to show that under appropriate conditions two homology theories are equivalent in a certain sense. We begin with the definition of a natural transformation between two homology theories A and B.

Definition: *Let A and B be homology theories. A **natural transformation** τ assigns to each space X homomorphisms $\tau : A_n(X) \to B_n(X)$ such that for each continuous map $f : X \to Y$ the diagram*

$$\begin{array}{ccc} A_n(X) & \xrightarrow{\tau} & B_n(X) \\ \downarrow f_* & & \downarrow f_* \\ A_n(Y) & \xrightarrow{\tau} & B_n(Y) \end{array}$$

commutes.

We furthermore require that the diagram

$$\begin{array}{ccc} A_n(U \cup V) & \xrightarrow{\tau} & B_n(U \cup V) \\ \downarrow d_A & & \downarrow d_B \\ A_{n-1}(U \cap V) & \xrightarrow{\tau} & B_{n-1}(U \cap V) \end{array}$$

commutes, where d_A and d_B are the boundary operators.

*A natural transformation is called a **natural equivalence** if for each X the homomorphisms $\tau : A_n(X) \to B_n(X)$ are isomorphisms.*

In the following chapters, we will sometimes consider two homology theories and a natural transformation between them, and we may want to check whether this is a natural equivalence — at least for a suitable class of spaces. It turns out that this can very easily be decided for the spaces we consider: one only has to check that $\tau : A_n(\text{pt}) \to B_n(\text{pt})$ is an isomorphism for all n.

To characterize such a class of suitable spaces, we introduce the notion of homology with compact supports. A space is called quasicompact if each open covering has a finite subcovering. If the space is also Hausdorff, then is is called compact.

Definition: *A homology theory h is a **homology theory with compact supports** (also called a **compactly supported homology theory**) if for each homology class $x \in h_n(X)$ there is a compact subspace $K \subset X$ and $\beta \in h_n(K)$ such that $x = j_*(\beta)$, where $j : K \to X$ is the inclusion, and if for each compact $K \subset X$ and $x \in h_n(K)$ mapping to 0 in $h_n(X)$, there is a compact space K' with $K \subset K' \subset X$ such that $i_*(x) = 0$, where $i : K \to K'$ is the inclusion.*

For example, integral homology and $\mathbb{Z}/2$-homology are theories with compact supports since the image of a quasicompact space under a continuous map is quasicompact.

A first comparison result is the following:

Proposition 9.1. *Let h and h' be homology theories with compact supports and let $\tau : h \to h'$ be a natural transformation such that $\tau : h_n(\text{pt}) \to h'_n(\text{pt})$ is an isomorphism for all n. Then τ is an isomorphism $\tau : h_n(U) \to h'_n(U)$ for all open $U \subset \mathbb{R}^k$.*

The proof is based on the **5-Lemma** from homological algebra which we now recall.

Lemma 9.2. *Consider a commutative diagram of abelian groups and homomorphisms*

$$\begin{array}{ccccccccc} A & \to & B & \to & C & \to & D & \to & E \\ \downarrow & & \downarrow \cong & & \downarrow f & & \downarrow \cong & & \downarrow \\ A' & \to & B' & \to & C' & \to & D' & \to & E' \end{array}$$

where the horizontal lines are exact sequences, the maps from B and D are isomorphisms, the map from A is surjective and the map from E is injective. Then the map $f : C \to C'$ is an isomorphism.

Proof: This is a simple diagram chasing argument. We demonstrate the principle by showing that $C \to C'$ is surjective and leave the injectivity as an exercise to the reader. For $c' \in C'$ consider the image $d' \in D'$ and the pre-image $d \in D$. Since E injects into E', the element d maps to 0 in E, and thus there is $c \in C$ mapping to d. By construction $f(c) - c'$ maps to 0 in D'. Thus there is $b' \in B'$ mapping to $f(c) - c'$. We take the pre-image $b \in B$ and replace c by $c - g(b)$, where g is the map from B to C. Then $f(c-g(b))-c' = f(c)-fg(b)-c' = f(c)-g'(b')-c' = f(c)-f(c)+c'-c' = 0$, where g' is the map from B' to C'.
q.e.d.

With this lemma we can now prove the proposition.

Proof of Proposition 9.1: Let U_1 and U_2 be open subsets of a space X and suppose that τ is an isomorphism for U_1, U_2 and $U_1 \cap U_2$. Then the Mayer-Vietoris sequence together with the 5-Lemma imply that τ is an isomorphism for $U_1 \cup U_2$.

Now consider a finite union of s open cubes $(a_1, b_1) \times \cdots \times (a_k, b_k) \subset \mathbb{R}^k$. Since the intersection of two open cubes is again an open cube or empty, the intersection of the s-th cube U_s with $U_1 \cup \cdots \cup U_{s-1}$ is a union of at most $s - 1$ open cubes. Since each cube is homotopy equivalent to a point pt, we conclude inductively over s that τ is an isomorphism for all $U \subset \mathbb{R}^k$ which are a finite union of s cubes.

Now consider an arbitrary $U \subset \mathbb{R}^k$ and $x \in h'_n(U)$. Since h'_n has compact supports, there is a compact subspace $K \subset U$ such that $x = j_*(\beta)$ with $\beta \in h'_n(K)$. Cover K by a finite union V of open cubes such that $K \subset V \subset U$ and denote the inclusion from K to V by i. Then, by the consideration above, $i_*(\beta)$ is in the image of $\tau : h_n(V) \to h'_n(V)$. Now consider the inclusion from V to U to conclude that x is in the image of

$\tau : h_n(U) \to h'_n(U)$. Thus τ is surjective.

For injectivity, one argues similarly. Let $x \in h_n(U)$ such that $\tau(x) = 0$. Then we first consider a compact subspace K in U such that $x = j_*(\beta)$ with $\beta \in h_n(K)$. Then, since $j_*(\tau(\beta)) = 0$ in $h'_n(U)$, there is a compact set K' such that $K \subset K' \subset U$ and $\tau(\beta)$ maps to 0 in $h'_n(K')$. By covering K' with a finite number of cubes in U, we conclude that β maps to 0 in this finite number of cubes since τ is injective for this space. Thus $x = 0$.
q.e.d.

Applying the Mayer-Vietoris sequence and the 5-Lemma again, one concludes that in the situation of Proposition 9.1 one can replace U by a space which can be covered by a finite union of open subsets which are homeomorphic to open subsets of $\mathbb{R}^k$.

Corollary 9.3. *Let h and h' be homology theories with compact supports and $\tau : h \to h'$ be a natural transformation. Suppose that $\tau : h_n(\text{pt}) \to h'_n(\text{pt})$ is an isomorphism for all n. Then for each topological manifold M (with or without boundary) admitting a finite atlas $\tau : h_n(M) \to h'_n(M)$ is an isomorphism for all n.*

In particular, this corollary applies to all compact manifolds. One can easily generalize this result by considering spaces $X = R \cup_f Y$ which are obtained by gluing a compact c-manifold R (i.e., a manifold together with a germ class of collars) via a continuous map $f : \partial R \to Y$ to a space Y for which τ is an isomorphism. For then we decompose $R \cup_f Y$ into $U := R - \partial R$ and V, the union of Y with the collar of ∂R in R. Then U is a manifold with finite atlas, $U \cap V$ is homotopy equivalent to ∂R, a manifold with finite atlas and V is homotopy equivalent to Y. Thus the corollary above together with the Mayer-Vietoris sequence and the 5-Lemma argument imply that τ is an isomorphism $h_n(R \cup_f Y) \to h'_n(R \cup_f Y)$.

Definition: *We call a space X **nice** if it is either a topological manifold (with or without boundary) with finite atlas or obtained by gluing a compact topological manifold with boundary to a nice space via a continuous map of the boundary.*

Corollary 9.4. *Let h, h' and τ be as above. Then for each nice space X and for all n the homomorphism $\tau : h_n(X) \to h'_n(X)$ is an isomorphism.*

3. CW-complexes

Motivated by the definition of nice spaces, we now introduce another class of objects called (finite) CW-complexes which lead to nice spaces. Of course, CW-complexes are useful in many aspects of algebraic topology aside from comparing homology theories.

Definition: *An m-dimensional finite CW-complex is a topological space X together with subspaces $\emptyset = X^{-1} \subset X^0 \subset X^1 \subset \cdots \subset X^m = X$. In addition we require that for $0 \leq j \leq m$, there are continuous maps $f_r^j = S_r^{j-1} \longrightarrow X^{j-1}$ and homeomorphisms*

$$\phi_j : (\bigsqcup_{r=1}^{s_j} D_r^j) \cup_{(\sqcup f_r^j)} X^{j-1} \cong X^j$$

*where $D_r^j = D^j = \{x \in \mathbb{R}^j \mid ||x|| \leq 1\}$, $S^{j-1} = \partial D_r^j$ and s_j is a non-negative integer (if $s_j = 0$ then we set $X^j = X^{j-1}$). We call X^0, X^1, ..., X^m a **CW-decomposition** of the topological space X and we call the subspaces $\phi_j(B_r^j)$ the j-cells of X. We denote a CW-complex simply by X.*

We see that a finite m-dimensional CW-complex can be obtained from a finite set of points with discrete topology by first attaching a finite number of 1-cells, followed by a finite number of 2-cells, ..., and finally a finite number of m-cells. All the k-cells are attached via continuous maps from their boundaries to X^{k-1}, the space already constructed from the cells of dimension less than or equal to $(k-1)$.

Remark: One can generalize the definition to non-finite CW-complexes which are obtained from an arbitrary discrete set by attaching an arbitrary number of 1-cells, 2-cells and so on.

Examples:
1) $X = S^m$, $X^0 = \cdots = X^{m-1} = \text{pt}$, $X^m = D^m \cup \text{pt}$.

2) Let $f^j : S^{j-1} \longrightarrow \mathbb{RP}^{j-1}$ be the canonical projection. Then we have a homeomorphism

$$D^j \cup_{f^j} \mathbb{RP}^{j-1} \longrightarrow \mathbb{RP}^j$$

mapping $x \in D^j$ to $[x_1, \ldots, x_j, \sqrt{1 - \Sigma x_j^2}]$ and $[x] \in \mathbb{RP}^{j-1}$ to $[x, 0]$. Thus $X^j := \mathbb{RP}^j$ ($0 \leq j \leq m$) gives a CW-decomposition of $\mathbb{RP}^m$.

3) Similarly, $X^j := \mathbb{C}\mathbb{P}^{[j/2]}$ gives a CW-decomposition of $\mathbb{C}\mathbb{P}^n$.

Here is a first instance showing that it is useful to consider CW-decompositions.

Theorem 9.5. *A finite CW-complex X is homologically and $\mathbb{Z}/2$-homologically finite. Denote the number of j-cells of a finite CW-complex X by β_j. Then:*

$$e(X) = \sum_{j=0}^{m}(-1)^j \cdot \beta_j.$$

Proof: We prove the statement inductively over the cells. Suppose that Y is homologically finite and $\mathbb{Z}/2$-homologically finite. Let $f : S^{k-1} \to Y$ be a continuous map and consider $Z := D^k \cup_f Y$. We decompose $Z = U \cup V$ with $U = \overset{\circ}{D^k}$ and $V = Z - \{0\}$, where $0 \in D^k$. The space $U \cap V$ is homotopy equivalent to S^{k-1}, U is homotopy equivalent to a point, and V is homotopy equivalent to Y.

The Mayer-Vietoris sequence implies that Z is homologically and $\mathbb{Z}/2$-homologically finite, thus, by Theorem 7.4,

$$\begin{aligned} e(Z) &= e(Y) + e(\text{pt}) - e(S^{k-1}) \\ &= e(Y) + 1 - (1 + (-1)^{k-1}) \\ &= e(Y) + (-1)^k, \end{aligned}$$

which implies the statement.
q.e.d.

Remark: In this case as well as in many other instances, it is enough to require that X is homotopy equivalent to a finite CW-complex.

4. Exercises

(1) Let h be a homology theory. Prove the following:
   a) If $f : A \to B$ is a homotopy equivalence then f_* is an isomorphism.
   b) For every two topological spaces there is a natural map $h_n(A) \oplus h_n(B) \to h_n(A \sqcup B)$ and it is an isomorphism.
   c) $h_n(\emptyset) = 0$ for all n.

(2) Let h be a homology theory.
   a) If $h_n(\text{pt}) = 0$ for all n, what can you say about h?

b) If there exists a non-empty space with $h_n(X) = 0$ for all n, what can you say about h?

(3) Which of the following are homology theories? Prove or disprove:
a) Given a topological space A define for every topological space X the homology groups $h_n(X) = SH_n(X \times A)$ and for a map $f : X \to Y$ the homomorphism $SH_n(X \times A) \to SH_n(Y \times A)$ induced by the map $f \times \mathrm{id} : X \times A \to Y \times A$.
b) Given a topological space A define for every topological space X the homology groups $h_n(X) = SH_n(X \sqcup A)$ and for a map $f : X \to Y$ the homomorphism $SH_n(X \sqcup A) \to SH_n(Y \sqcup A)$ induced by the map $f \sqcup \mathrm{id} : X \sqcup A \to Y \sqcup A$.
c) Define for every topological space X the homology groups $h_n(X) = SH_n(X \times X)$ and for a map $f : X \to Y$ the homomorphism $SH_n(X \times X) \to SH_n(Y \times Y)$ induced by the map $f \times f : X \times X \to Y \times Y$.

(4) Define for every topological space a series of functors $h_n(X) = SH_n(X) \otimes \mathbb{Z}/2$ with the boundary operator $d \otimes \mathbb{Z}/2$.
a) Show that there is a natural transformation
$$\eta : h_n(X) \to SH_n(X; \mathbb{Z}/2).$$
b) Is η a natural isomorphism?
c) Is h a homology theory?

(5) Let h and h' be two homology theories. Show that $h \oplus h'$ is a homology theory.

(6) Let h be a homology theory. Show that h' defined by $h'_n(X) = h_{n+k}(X)$ for a given k is a homology theory.

(7) Let X be a topological space and $f_1 : M_1 \to X$, $f_2 : M_2 \to X$ be two continuous maps from closed manifolds of dimension n. We say that the two maps are bordant if there is a compact manifold M with boundary equal to $M_1 \sqcup M_2$ and a map $f : M \to X$ such that $f|_{M_i} = f_i$. Define $\mathbf{N}_n(X)$ to be the set of bordism classes of maps $f : M^n \to X$ where M^n is a closed manifold of dimension n and f is a continuous map. Show that $\mathbf{N}_*$ is a homology theory with the boundary operator d defined in a similar way to the one we defined for SH_n. Show that $\mathbf{N}_1(\mathrm{pt})$ is trivial and $\mathbf{N}_2(\mathrm{pt})$ is generated by $\mathbb{RP}^2$.

(8) a) Define $SH_n^p(X)$ in a similar way to the way we defined $SH_n(X)$, but instead of stratifolds we use p-stratifolds and the same for stratifolds with boundary. Show that this is a homology theory. What can you say about its connection to $SH_n(X)$?

4. Exercises

b) Show that every class in $SH_n^p(X)$ for $n \leq 2$ can be represented by a map from a stratifold which is actually a manifold.

Chapter 10

Künneth's theorem

Prerequisites: in this chapter we assume that the reader is familiar with tensor products of modules. The basic definitions and some results on tensor products relevant to our context are contained in Appendix C.

1. The cross product

We want to compute the homology of $X \times Y$. To compare it with the homology of X and Y, we construct the $\times$-product $SH_i(X) \times SH_j(Y) \to SH_{i+j}(X \times Y)$. If $[\mathbf{S}, g] \in SH_k(X)$ and $[\mathbf{S}', g'] \in SH_\ell(Y)$ we construct an element

$$[\mathbf{S}, g] \times [\mathbf{S}', g'] \in SH_{k+\ell}(X \times Y)$$

and similarly for $\mathbb{Z}/2$-homology.

For this we take the cartesian product of $\mathbf{S}$ and $\mathbf{S}'$ (considered as a stratifold by example 6 in chapter 2) and the product of g and g'.

If $\mathbf{S}$ and $\mathbf{S}'$ are regular of dimension k and l and $\mathbb{Z}/2$-oriented, then the product is regular and the $(k+\ell-1)$-dimensional stratum $\bigsqcup_{i+j=k+\ell-1}(\mathbf{S}^i \times (\mathbf{S}')^j) = (\mathbf{S}^k \times (\mathbf{S}')^{\ell-1}) \sqcup (\mathbf{S}^{k-1} \times (\mathbf{S}')^\ell)$ is empty which means that $\mathbf{S} \times \mathbf{S}'$ is also $\mathbb{Z}/2$-oriented. Thus $[\mathbf{S} \times \mathbf{S}', g \times g']$ is an element of $SH_{k+\ell}(X \times Y; \mathbb{Z}/2)$. If $\mathbf{S}$ and $\mathbf{S}'$ are oriented then the $(k+\ell)$-dimensional stratum is $\mathbf{S}^k \times (\mathbf{S}')^\ell$ and carries the product orientation. Thus $[\mathbf{S} \times \mathbf{S}', g \times g']$ is an element of $SH_{k+\ell}(X \times Y)$. This is the construction of the $\times$-**products or cross**

products :
$$SH_i(X) \times SH_j(Y) \to SH_{i+j}(X \times Y)$$
and
$$SH_i(X; \mathbb{Z}/2) \times SH_j(X; \mathbb{Z}/2) \to SH_{i+j}(X \times Y; \mathbb{Z}/2)$$
which are defined as
$$[\mathbf{S}, g] \times [\mathbf{S}', g'] := [\mathbf{S} \times \mathbf{S}', g \times g'].$$

The following Proposition follows from the definition of the $\times$-product:

Proposition 10.1. *The $\times$-products are bilinear and associative.*

Since the $\times$-products are bilinear they induce maps from the tensor product
$$SH_i(X; \mathbb{Z}/2) \otimes_{\mathbb{Z}/2} SH_j(Y; \mathbb{Z}/2) \longrightarrow SH_{i+j}(X \times Y; \mathbb{Z}/2)$$
and
$$SH_i(X) \otimes SH_j(Y) \longrightarrow SH_{i+j}(X \times Y).$$

(We denote the tensor product of abelian groups by $\otimes$ and of F-vector spaces by $\otimes_F$.)

We sum the left side over all i, j with $i+j = k$ to obtain homomorphisms
$$\times : \bigoplus_{i+j=k} SH_i(X) \otimes SH_j(Y) \to SH_k(X \times Y)$$
and
$$\times : \bigoplus_{i+j=k} SH_i(X; \mathbb{Z}/2) \otimes_{\mathbb{Z}/2} SH_j(Y; \mathbb{Z}/2) \to SH_k(X \times Y; \mathbb{Z}/2).$$

It would be nice if these maps were isomorphisms. For $\mathbb{Z}/2$-homology, we will show this under some assumptions on X, but for integral homology these assumptions are not sufficient. The idea is to fix Y and to consider the functor
$$SH_k^Y(X) := SH_k(X \times Y)$$
where for $f : X \to X'$ we define
$$f_* : H_k^Y(X) \to H_k^Y(X')$$

1. The cross product

by $(f \times \mathrm{id})_*$. This is obviously a homology theory: the Mayer-Vietoris sequence holds since
$$(U_1 \times Y) \cup (U_2 \times Y) = (U_1 \cup U_2) \times Y,$$
$$(U_1 \times Y) \cap (U_2 \times Y) = (U_1 \cap U_2) \times Y.$$
Furthermore this is a homology theory with compact supports.

For X a point the maps $\times$ above are isomorphisms. Thus we could try to prove that they are always an isomorphism for nice spaces X by applying the comparison result Corollary 9.4 if
$$X \longmapsto \bigoplus_{i+j=k} SH_i(X) \otimes SH_j(Y) =: h_k^Y(X)$$
were also a homology theory and, similarly, if
$$X \longmapsto \bigoplus_{i+j=k} SH_i(X; \mathbb{Z}/2) \otimes_{\mathbb{Z}/2} SH_j(Y; \mathbb{Z}/2) =: h_k^Y(X; \mathbb{Z}/2)$$
were a homology theory. Here, for $f : X \to X'$, we define
$$f_* = \bigoplus_{i+j=k} ((f_* \otimes \mathrm{id}) : SH_i(X) \otimes SH_j(Y) \to SH_i(X') \otimes SH_j(Y)).$$

The homotopy axiom is clear but the Mayer-Vietoris sequence is a problem. It would follow if for an exact sequence of abelian groups
$$A \xrightarrow{f} B \xrightarrow{g} C$$
and an abelian group D the sequence
$$A \otimes D \xrightarrow{f \otimes \mathrm{id}} B \otimes D \xrightarrow{g \otimes \mathrm{id}} C \otimes D$$
were exact. But this is in general not the case. For example consider
$$0 \to \mathbb{Z} \xrightarrow{\cdot 2} \mathbb{Z}$$
and $D = \mathbb{Z}/2$ giving
$$0 \to \mathbb{Z}/2\mathbb{Z} \xrightarrow{\cdot 2} \mathbb{Z}/2\mathbb{Z}$$
which is not exact since $\cdot 2 : \mathbb{Z}/2\mathbb{Z} \to \mathbb{Z}/2\mathbb{Z}$ is 0. If instead of abelian groups we work with vector spaces over a field F, the sequence
$$A \otimes_F D \xrightarrow{f \otimes_F \mathrm{id}} B \otimes_F D \xrightarrow{g \otimes_F \mathrm{id}} C \otimes_F D$$
is exact. It is enough to show this for short exact sequences $0 \to A \to B \to C \to 0$ by passing to the image in C and dividing out the kernel in A. Then

there is a splitting $s : C \to B$ with $gs = \text{id}$ and a splitting $p : B \to A$ with $pf = \text{id}$. These splittings induce splittings of

$$A \otimes_F D \xrightarrow{f \otimes_F \text{id}} B \otimes_F D \xrightarrow{g \otimes_F \text{id}} C \otimes_F D,$$

implying its exactness.

The sequence is also exact if D is a torsion-free finitely generated abelian group. Namely then $D \cong \mathbb{Z}^r$ for some r. It is enough to check exactness for $r = 1$, where it is trivial since $A \otimes \mathbb{Z} \cong A$. For larger r we use that $A \otimes (D \oplus D') \cong (A \otimes D) \oplus (A \otimes D')$.

Thus, if all homology groups of Y are finitely generated and torsion-free, the functor $h_k^Y(X)$ is a generalized homology theory. And since $SH_k(X; \mathbb{Z}/2)$ is a $\mathbb{Z}/2$-vector space we conclude that for any fixed space Y the functor $h_k^Y(X; \mathbb{Z}/2)$ is a homology theory. To obtain some partial information about the integral homology groups of a product of two spaces, if $SH_k(Y)$ is not finitely generated and torsion-free, we define rational homology groups.

Definition: $SH_m(X; \mathbb{Q}) := SH_m(X) \otimes \mathbb{Q}$. For $f : X \to Y$ we define $f_* : SH_m(X; \mathbb{Q}) \to SH_m(Y; \mathbb{Q})$ by $f_* \otimes \text{id} : SH_m(X) \otimes \mathbb{Q} \to SH_m(Y) \otimes \mathbb{Q}$.

By the considerations above the rational homology groups define a homology theory called **rational homology**. Since the rational homology groups are $\mathbb{Q}$-vector spaces (scalar multiplication with $\lambda \in \mathbb{Q}$ is given by $\lambda(x \otimes \mu) := x \otimes \lambda\mu$), the functor

$$X \longmapsto \bigoplus_{i+j=k} SH_i(X; \mathbb{Q}) \otimes SH_j(Y; \mathbb{Q}) =: h_k^Y(X; \mathbb{Q})$$

is a homology theory. By construction it has compact supports.

2. The Künneth theorem

To apply Corollary 9.4 we have to check that the maps $\times : h_k^Y(X; \mathbb{Z}/2) \to SH_k^Y(X; \mathbb{Z}/2)$, $\times : h_k^Y(X) \to SH_k^Y(X)$ and $\times : h_k^Y(X; \mathbb{Q}) \to SH_k^Y(X; \mathbb{Q})$ commute with induced maps and the boundary operator in the Mayer-Vietoris sequence, in other words, that these maps are natural transformations. The proof is the same in both cases and so we only give it for $\mathbb{Z}/2$-homology:

Lemma 10.2. *The maps*
$$\times : h_k^Y(X; \mathbb{Z}/2) \to SH_k^Y(X; \mathbb{Z}/2)$$
define a natural transformation.

Proof: Everything is clear except the commutativity in the Mayer-Vietoris sequences. Let U_1 and U_2 be open subsets of X and consider for $i+j = k$ the element $[\mathbf{S}, f] \otimes [\mathbf{Z}, g] \in SH_i(U_1 \cup U_2) \otimes SH_j(Y)$. By our definition of the boundary operator in the Mayer-Vietoris sequence of $SH_i(X)$ we can decompose the stratifold $\mathbf{S}$ (after perhaps changing it by a bordism) as $\mathbf{S} = \mathbf{S}_1 \cup \mathbf{S}_2$ with $\partial \mathbf{S}_1 = \partial \mathbf{S}_2 =: \mathbf{Q}$, where $f(\mathbf{S}_1) \subset U_1$ and $f(\mathbf{S}_2) \subset U_2$. Then $d([\mathbf{S}, f]) = [\mathbf{Q}, f|_\mathbf{Q}]$. Thus $d([\mathbf{S}, f] \otimes [\mathbf{Z}, g]) = [\mathbf{Q}, f|_\mathbf{Q}] \otimes [\mathbf{Z}, g]$. On the other hand $[\mathbf{S}, f] \times [\mathbf{Z}, g] = [\mathbf{S} \times \mathbf{Z}, f \times g]$ and, since $\mathbf{S} \times \mathbf{Z} = (\mathbf{S}_1 \cup \mathbf{S}_2) \times \mathbf{Z}$, we conclude: $d([\mathbf{S} \times \mathbf{Z}, f \times g]) = [\mathbf{Q} \times \mathbf{Z}, f|_\mathbf{T} \times g]$. Thus the diagram

$$\begin{array}{ccc} SH_i(U_1 \cup U_2) \otimes SH_j(Y) & \xrightarrow{\times} & SH_{i+j}((U_1 \cup U_2) \times Y) \\ \downarrow d & & \downarrow d \\ SH_{i-1}(U_1 \cap U_2) \otimes SH_j(Y) & \xrightarrow{\times} & SH_{i+j-1}((U_1 \cap U_2) \times Y) \end{array}$$

commutes.
q.e.d.

Now the Künneth Theorem is an immediate consequence of Corollary 9.4:

Theorem 10.3. (Künneth Theorem) *Let X be a nice space (see page 95). Then for $F = \mathbb{Q}$ or $\mathbb{Z}/2\mathbb{Z}$*
$$\times : \bigoplus_{i+j=k} SH_i(X; F) \otimes_F SH_j(Y; F) \to SH_k(X \times Y; F)$$
is an isomorphism. The same holds for integral homology if for all j the groups $SH_j(Y)$ are torsion-free and finitely generated.

We note that the Kunneth theorem stated here holds for all spaces X which admit decompositions as finite CW-complexes since all these spaces are nice. In a later chapter we will identify the stratifold homology of CW-complexes with the homology groups defined in a traditional way using simplices. The world of chain complexes is more appropriate for dealing with the Künneth Theorem and there one obtains a general result computing the integral homology groups of a product of CW-complexes.

As an application we prove that for nice spaces X the Euler characteristic of $X \times Y$ is the product of the Euler characteristics of X and Y.

Theorem 10.4. *Let X and Y be $\mathbb{Z}/2$ homologically finite and X a nice space. Then*
$$e(X \times Y) = e(X) \cdot e(Y).$$

Proof: By the previous theorem the proof follows from

Lemma 10.5. *Let $A = (A_0, A_1, \ldots, A_k)$ and $B = (B_0, \ldots, B_r)$ be sequences of finite-dimensional $\mathbb{Z}/2$-vector spaces. Then for $C = C(A, B) = (C_0, C_1, \ldots, C_{k+r})$ with $C_s := \bigoplus_{i+j=s} A_i \otimes B_j$, we have*
$$e(C) = e(A) \cdot e(B)$$
where $e(A) := \sum_i (-1)^i \dim A_i$ and similarly for $e(B)$ and $e(C)$.

Proof: The proof is by induction over k and for $k = 0$ the result is clear. Let A' be given by $A_0, A_1, \ldots, A_{k-1}$. Then $e(A') + (-1)^k \dim A_k = e(A)$. Define C' as $C(A', B)$. Then $C_s = C'_s$ for $s \leq k$ and $C_{k+j} = C'_{k+j} \oplus (A_k \otimes B_j)$. Thus,
$$\begin{aligned} e(C) &= e(C') + (-1)^k \dim A_k \, e(B) \\ &= e(A') \cdot e(B) + (-1)^k \dim A_k \, e(B) \\ &= e(A) \cdot e(B). \end{aligned}$$

q.e.d.

Another application is the computation of the homology of a product of two spheres $S^n \times S^m$ for n and m positive. Since the homology groups of S^m are torsion-free, the Künneth theorem implies $H_k(S^n \times S^m) = \bigoplus_{i+j=k} H_i(S^n) \otimes H_j(S^m)$. Using the fact that $A \otimes \mathbb{Z} \cong A$ for each abelian group A we obtain

$$H_k(S^n \times S^m) \cong \begin{cases} \mathbb{Z} & k = 0, n+m \\ \mathbb{Z} & k = n, \text{if } n \neq m \\ \mathbb{Z} & k = m, \text{if } n \neq m \\ \mathbb{Z} \oplus \mathbb{Z} & k = n = m \\ 0 & \text{otherwise.} \end{cases}$$

With the $\times$-product we also obtain a basis for the homology groups of $S^n \times S^m$. Let x be a point in S^n and y be a point in S^m. Then $[(x,y), i]$ generates $H_0(S^n \times S^m)$, $[S^n \times y, i]$ generates $H_n(S^n \times S^m)$ and $[x \times S^m, i]$ generates $H_m(S^n \times S^m)$ for $n \neq m$, and these elements are a basis of $H_n(S^n \times S^n)$, if $n = m$, and finally the fundamental class $[S^n \times S^m]$ generates $H_{n+m}(S^n \times S^m)$. Here i always stands for the inclusion.

These examples agree with our geometric intuition that the manifolds giving the homology classes "catch" the corresponding holes.

3. Exercises

(1) Compute the homology of the following spaces. Can you represent its elements by maps from stratifolds?
$S^1 \times S^1$, or more generally $T^k = S^1 \times S^1 \times \cdots \times S^1$, the product of k copies of S^1.

(2) Let X be a homologically finite space. Denote by $P(X) = \sum a_k x^k$ the polynomial with $a_k = \mathrm{rank}(SH_k(X))$.
a) Show that for homologically finite spaces $P(X \times Y) = P(X) \times P(Y)$.
b) Compute $P(S^n)$ and conclude that S^n is not homeomorphic to the product of two manifolds of positive dimensions.

(3) Compute the integral homology of $\mathbb{RP}^2 \times \mathbb{RP}^2$. Does the Künneth formula hold? Can you explain this?

Chapter 11

Some lens spaces and quaternionic generalizations

1. Lens spaces

In this chapter we will construct a class of manifolds that, on the one hand, gives more fundamental examples to play with and, on the other hand, is the basis for some very interesting aspects of modern differential topology. Some of these aspects will be discussed in later chapters.

The manifolds under consideration have various geometric features. We will concentrate on one aspect: they are total spaces of smooth fibre bundles. A **smooth fibre bundle** is a smooth map $p : E \to B$ between smooth manifolds such that for each $x \in B$ there are: an open neighbourhood U of x, a smooth manifold F and a diffeomorphism $\varphi : p^{-1}(U) \to U \times F$ with $p|_{p^{-1}(U)} = p_1 \varphi$ where p_1 is the projection from $U \times F$ onto U. Such a φ is called a **local trivialization** of p. For a point $x \in B$ we call $E_x := p^{-1}(x)$ the **fibre** over x. Observe that $\varphi|E_x$ defines a diffeomorphism from E_x to F.

We begin with some bundles over S^2 with fibre S^1. Let k be an integer. Decompose S^2 as $D^2 \cup_{S^1} D^2$ and define

$$L_k := D^2 \times S^1 \cup_{f_k} D^2 \times S^1$$

where $f_k : S^1 \times S^1 \to S^1 \times S^1$ is the diffeomorphism mapping (z_1, z_2) to $(z_1, z_1^k z_2)$. Here we consider S^1 as a subgroup of $\mathbb{C}^*$. The map is a diffeomorphism since $(z_1, z_2) \mapsto (z_1, z_1^{-k} z_2)$ is the inverse map. L_k is equipped with a smooth structure. It is called a **lens space**. Is it orientable? This is easily seen without deeper consideration for the following reason. Since $S^1 \times S^1 = \partial(D^2 \times S^1)$ is connected, f_k is either orientation-preserving or orientation-reversing (by continuity of the orientation and of df_x the orientation behaviour cannot jump). If it were orientation-reversing, we are done by orienting both copies of $D^2 \times S^1$ in $D^2 \times S^1 \cup_f D^2 \times S^1$ equally. If it is orientation-preserving, we are also done by orienting the second copy of $D^2 \times S^1$ in $D^2 \times S^1 \cup_f D^2 \times S^1$ opposite to the first one, making f_k artificially orientation-reversing.

Of course, by computing $d(f_k)_x$ in our example we can decide if f_k is orientation-preserving. For this consider $S^1 \times S^1$ as a submanifold of $\mathbb{C}^* \times \mathbb{C}^*$ and extend f_k to the map given by the same expression on $\mathbb{C}^* \times \mathbb{C}^*$. Then $(df_k)_{(z_1, z_2)}$ is given by the complex Jacobi matrix

$$\begin{pmatrix} 1 & 0 \\ k z_1^{k-1} z_2 & z_1^k \end{pmatrix}.$$

To obtain the map on $T_{(z_1, z_2)}(S^1 \times S^1)$ we have to restrict this map to $z_1^\perp \times z_2^\perp = T_{(z_1, z_2)}(S^1 \times S^1)$. We give a basis of $T_{(z_1, z_2)}(S^1 \times S^1)$ by $(iz_1, 0)$ and $(0, iz_2)$ and use this basis as our standard orientation. We have to compare the orientation given by $d(f_k)_{(z_1, z_2)}(iz_1, 0)$ and $d(f_k)_{(z_1, z_2)}(0, iz_2)$ at the point $f_k(z_1, z_2) = (z_1, z_1^k z_2)$ with that given by $(iz_1, 0)$ and $(0, iz_1^k z_2)$. But $d(f_k)_{(z_1, z_2)}(iz_1, 0) = (iz_1, kiz_1^k z_2)$ and $d(f_k)_{(z_1, z_2)}(0, iz_2) = (0, iz_1^k z_2)$. The change of basis matrix is

$$\begin{pmatrix} 1 & 0 \\ k & 1 \end{pmatrix}$$

and it has a positive determinant. Thus f_k is orientation-preserving and to orient L_k we have to consider it as $D^2 \times S^1 \cup_{f_k} -D^2 \times S^1$. From now on, we consider L_k as an oriented 3-manifold with this orientation.

There are different natural descriptions of lens spaces. Although we don't need it, we shall give another description of L_k for $k > 0$. For this we consider the 3-sphere S^3 as the subspace of $\mathbb{C}^2$ consisting of pairs (v_1, v_2) with $|v_1| + |v_2| = 1$ where $|v_i|$ is the norm of the complex number v_i, $i = 1$ or 2. The group of k-th roots of unity in S^1 is $G_k = \{z \in S^1 \mid z^k = 1\}$ and this acts on $\mathbb{C}^2$ by $z \cdot (v_1, v_2) = (zv_1, zv_2)$. Clearly this action preserves $S^3 \subset \mathbb{C}^2$. We consider the space $S^3/G_k := S^3/\sim$, where $(v_1, v_2) \sim (w_1, w_2)$ if and only

1. Lens spaces

if there is a $z \in G_k$ such that $z \cdot (v_1, v_2) = (w_1, w_2)$. For example, S^3/G_2 is the projective space $\mathbb{RP}^3$. It is not difficult to identify S^3/G_k with L_k. As a hint one should start with the case $k = 1$ and identify $S^3 = S^3/G_1$ with L_1. Once this is achieved, one can use this information to solve the case $k > 1$.

We consider the map $p : L_k \to S^2 = D^2 \cup -D^2$ mapping $(z_1, z_2) \in D^2 \times S^1$ to z_1 and $(z_1, z_2) \in -D^2 \times S^1$ to z_1. This is obviously well defined and by construction of the smooth structures on L_k and on $D^2 \cup -D^2 = S^2$ it is a smooth map. Actually, by construction $p : L_k \to S^2$ is a smooth fibre bundle.

We want to classify the manifolds L_k up to diffeomorphism. For this we first compute the homology groups. We prepare for this with some general considerations. As above, consider two smooth c-manifolds W_1 and W_2 and a diffeomorphism $f : \partial W_1 \to \partial W_2$. Then consider the open covering of $W_1 \cup_f W_2$ given by the union of W_1 and the collar of ∂W_2 in W_2, denoted by U, and of W_2 and the collar of ∂W_1 in W_1, denoted by V. Obviously, the inclusions from W_1 to U and from W_2 to V as well as from ∂W_1 to $U \cap V$ are homotopy equivalences. With this information we consider the Mayer-Vietoris sequence and replace the homology group of U, V and $U \cap V$ by the isomorphic homology group of W_1, W_2 and ∂W_1:

$$\cdots \to SH_k(\partial W_1) \to SH_k(W_1) \oplus SH_k(W_2)$$
$$\to SH_k(W_1 \cup_f W_2) \xrightarrow{d} SH_{k-1}(\partial W_1) \to \cdots$$

where the map from $SH_k(W_1) \oplus SH_k(W_2)$ to $SH_k(W_1 \cup W_2)$ is the difference of the maps induced by inclusions. The map from $SH_k(\partial W_1)$ to $SH_k(W_1)$ is $(j_1)_*$, where j_1 is the inclusion from ∂W_1 to W_1, and the map from $SH_k(\partial W_1)$ to $SH_k(W_2)$ is $(j_2)_* f_*$, where j_2 is the inclusion from ∂W_2 to W_2.

Applying this to L_k implies that $SH_r(L_k) = 0$ for $r > 3$ and we have an isomorphism $SH_3(L_k) \xrightarrow{d} SH_2(S^1 \times S^1) \cong \mathbb{Z}$. Since the fundamental class $[L_k] \in SH_3(L_k)$ is a primitive element, we conclude

$$SH_3(L_k) = \mathbb{Z}[L_k],$$

the free abelian group of rank 1 generated by the fundamental class $[L_k]$. The computation of SH_2 and SH_1 is given by the exact sequence:

$$0 \to SH_2(L_k) \to SH_1(S^1 \times S^1) \to SH_1(S^1) \oplus SH_1(S^1) \to SH_1(L_k) \to 0$$

in which the map from $SH_1(S^1 \times S^1)$ to the first component is $(p_2)_*$, where p_2 is the projection onto the second factor, and the map from $SH_1(S^1 \times S^1)$ to the second component is $(p_2)_*(f_k)_*$. By the Künneth Theorem 10.3 we have seen that $SH_1(S^1 \times S^1) = \mathbb{Z}[S^1, i_1] \oplus \mathbb{Z}[S^1, i_2]$, where $i_1(z) = (z, 1)$ and $i_2(z) = (1, z)$. If α is an element of $SH_1(S^1 \times S^1)$, the coefficients of α with respect to the basis $[S^1, i_1]$ and $[S^1, i_2]$ are $(p_1)_*(\alpha) \in SH_1(S^1) = \mathbb{Z}$ and $(p_2)_*(\alpha) \in SH_1(S^1) = \mathbb{Z}$.

Thus
$$(f_k)_*[S^1, i_1] = \deg(p_1 f_k i_1)[S^1, i_1] + \deg(p_2 f_k i_1)[S^1, i_2]$$
and
$$(f_k)_*[S^1, i_2] = \deg(p_1 f_k i_2)[S^1, i_1] + \deg(p_2 f_k i_2)[S^1, i_2].$$

From Proposition 8.4 we know the corresponding degrees and conclude that with respect to the basis $[S^1, i_1]$ and $[S^1, i_2]$ of $SH_1(S^1 \times S^1)$ the map $(f_k)_*$ is given by
$$\begin{pmatrix} 1 & 0 \\ k & 1 \end{pmatrix}.$$

With this information the exact sequence above gives
$$0 \to SH_2(L_k) \to \mathbb{Z} \oplus \mathbb{Z} \to \mathbb{Z} \oplus \mathbb{Z} \to SH_1(L_k) \to 0,$$
where the map $\mathbb{Z} \oplus \mathbb{Z} \to \mathbb{Z} \oplus \mathbb{Z}$ is given by the matrix
$$\begin{pmatrix} 0 & 1 \\ k & 1 \end{pmatrix}.$$

The kernel of this linear map is 0 and the cokernel $\mathbb{Z}/|k|\mathbb{Z}$ (exercise). Thus we have shown that $SH_2(L_k) = 0$ and $SH_1(L_k) \cong \mathbb{Z}/|k|\mathbb{Z}$ generated by $i_*[S^1]$ where $i: S^1 \to L_k$ is the inclusion of any fibre.

Proposition 11.1. *The homology of L_k is*
$$SH_r(L_k) \cong \begin{cases} 0 & r > 3, r = 2 \\ \mathbb{Z} & r = 0, 3 \\ \mathbb{Z}/|k|\mathbb{Z} & r = 1 \end{cases}$$

where $SH_1(L_k)$ is generated by $[S^1, j]$ and $j: S^1 \to D^2 \times S^1 \subset L_k$ maps z to $(0, z)$.

As a consequence, $|k|$ is an invariant of the homeomorphism type or even the homotopy type of L_k. On the other hand, we observe that the diffeomorphism $c: S^1 \times S^1 \to S^1 \times S^1$ which maps (z_1, z_2) to $(z_1, \bar{z}_2)$ where $\bar{z}_2$ is the conjugate of z_2, satisfies $c \circ f_k = f_{-k} \circ c$. It follows that we may construct a diffeomorphism from L_k to L_{-k} which is conjugation on each fibre and which on each $D^2 \times S^1$ has the form $(w_1, z_2) \mapsto (w_1, \bar{z}_2)$. Thus we conclude:

Proposition 11.2. L_k is diffeomorphic to L_q if and only if $|k| = |q|$.

2. Milnor's 7-dimensional manifolds

Now, we generalize our construction by passing from the complex numbers to the **quaternions** $\mathbb{H}$. Recall that $\mathbb{H}$ is the skew field which as an abelian group is $\mathbb{R}^4$ with basis $1, i, j, k$ and multiplication defined by the relations $i^2 = j^2 = k^2 = -1$ and $ij = -ji, ik = -ki, jk = -kj$ and $ij = k, jk = i, ki = j$. It is useful to consider $\mathbb{H}$ as $\mathbb{C} \times \mathbb{C}$ with $1 = (1,0), i = (i,0), j = (0,1)$ and $k = (0,i)$. Then the multiplication is given by the formula

$$(z_1, z_2) \cdot (y_1, y_2) = (z_1 y_1 - \bar{y}_2 z_2, y_2 z_1 + z_2 \bar{y}_1).$$

The unit vectors of $\mathbb{H} \cong \mathbb{C}^2$ give the 3-sphere, $S^3 = \{(z_1, z_2) \mid z_1 \bar{z}_1 + z_2 \bar{z}_2 = 1\}$, and form a multiplicative subgroup of $\mathbb{H}$. In contrast to $S^1 \subset \mathbb{C}$, this subgroup is not commutative. This is the reason why we have more possibilities when we generalize our construction of L_k to the quaternions.

Let k and ℓ be integers. Then we define a diffeomorphism

$$\begin{aligned} f_{k,\ell} : S^3 \times S^3 &\longrightarrow S^3 \times S^3 \\ (x, y) &\longmapsto (x, x^k y x^\ell) \end{aligned}$$

and define

$$M_{k,\ell} := D^4 \times S^3 \cup_{f_{k,\ell}} -D^4 \times S^3.$$

The map $f_{k,\ell}$ is a diffeomorphism since it has inverse $f_{-k,-\ell}$. As in the case of lens spaces, one can show that $f_{k,\ell}$ is orientation-preserving, and thus one has to take the opposite orientation on the second copy of $D^4 \times S^3$ to orient $M_{k,\ell}$ in a consistent way. As for lens spaces the projection onto $D^4 \cup -D^4$ gives a smooth fibre bundle $p : M_{k,\ell} \to S^4$. We call these manifolds **Milnor manifolds**, since they were investigated by Milnor in his famous paper "On manifolds homeomorphic to the 7-sphere" [**Mi 1**].

We can compute $SH_r(M_{k,\ell})$ in the same way as $SH_r(L_k)$ once we know the induced map

$$(f_{k,\ell})_* : SH_3(S^3 \times S^3) \to SH_3(S^3 \times S^3).$$

To compute this, consider two maps $f, g : S^3 \to S^3$. We compute the degree of

$$\begin{aligned} f \cdot g : S^3 &\longrightarrow S^3 \\ x &\longmapsto f(x) \cdot g(x). \end{aligned}$$

Lemma 11.3. *For continuous maps $f, g : S^3 \to S^3$ the degree of $f \cdot g$ is*

$$\deg(f \cdot g) = \deg f + \deg g.$$

Proof: Consider the diagonal map $\triangle : S^3 \to S^3 \times S^3$ mapping $x \mapsto (x, x)$. The map on homology induced by $\triangle$ maps the fundamental class $[S^3]$ to $[S^3, i_1] + [S^3, i_2]$, where $i_1(q) = (q, 1)$ and $i_2(q) = (1, q)$ (one can either construct a bordism between the two classes or use the Künneth theorem). The map $\mu : S^3 \times S^3 \to S^3$ sending (q_1, q_2) to $f(q_1) \cdot g(q_2)$ induces a map in homology mapping $[S^3, i_1]$ to $\deg f \cdot [S^3]$, and $[S^3, i_2]$ to $\deg g \cdot [S^3]$ (exercise). Thus since $f \cdot g = \mu \circ (f \times g)$, we see that $\deg f \cdot g = \deg f + \deg g$. q.e.d.

With this information one concludes that, with respect to the basis $[S^3, i_1]$ and $[S^3, i_2]$ of $SH_3(S^3 \times S^3)$, the induced map of $f_{k,\ell}$ on $SH_3(S^3 \times S^3)$ is given by the matrix

$$\begin{pmatrix} 1 & 0 \\ k+l & 1 \end{pmatrix}.$$

From this, as in the case of L_k, one can compute the homology of $M_{k,\ell}$ and obtains:

Proposition 11.4. $SH_r(M_{k,\ell}) = 0$ *for $r > 7$ and $r = 1, 2, 5$ and 6, whereas*

$$SH_0(M_{k,\ell}) = \mathbb{Z}$$

$$SH_7(M_{k,\ell}) = \mathbb{Z} \cdot [M_{k,\ell}]$$

$$SH_3(M_{k,\ell}) \cong \mathbb{Z}/|k+\ell| \cdot \mathbb{Z}$$

$$SH_4(M_{k,\ell}) = \begin{cases} \mathbb{Z} & k+\ell = 0 \\ 0 & \text{otherwise.} \end{cases}$$

Thus $|k+\ell|$ is an invariant of the homotopy type. In contrast to L_k, this is not enough to distinguish the manifolds $M_{k,\ell}$. In the next chapters we will develop various techniques of general interest which will have surprising implications for the manifolds $M_{k,\ell}$. In fact, we shall see that these manifolds serve as wonderful motivating examples which illustrate highly important theories concerning the structure of manifolds.

3. Exercises

(1) Show that the diffeomorphism type of fibres of a smooth fibre bundle doesn't change on connected components.

(2) Let $p : E \to B$ be a smooth fibre bundle. Show that if $\dim E = \dim B$ and B is connected a smooth fibre bundle is a covering and that a covering with a countable number of sheets by a smooth map is a smooth fibre bundle.

(3) For a smooth manifold M and a self-diffeomorphism f consider the mapping torus M_f. Show that $p_2 : M_f \to [0,1] = S^1$ is a smooth fibre bundle.

(4) A map $p : E \to M$ between smooth manifolds is called proper if the preimage of each compact subset is compact. **Ehresmann's theorem** says that a proper smooth submersion $p : E \to M$ (i.e., all points in M are regular values) is a smooth fibre bundle.
Is the condition *proper* needed?

(5) a) Show that the **Klein bottle** $S^1 \times S^1 /_\tau$, where $\tau(x,y) = (-x, \bar{y})$ is the total space of a smooth fibre bundle over S^1. Determine its fibre.
b) Show that it is homeomorphic to the connected sum $\mathbb{RP}^2 \# \mathbb{RP}^2$.
c) Show that this bundle is non-trivial, i.e., not isomorphic to the product bundle.

(6) Show that L_k is diffeomorphic to S^3/G_k with the action as explained in this chapter.

(7) Show that the cokernel of the map $\mathbb{Z} \oplus \mathbb{Z} \to \mathbb{Z} \oplus \mathbb{Z}$ given by the matrix
$$\begin{pmatrix} 0 & 1 \\ k & 1 \end{pmatrix}$$
is $\mathbb{Z}/|k|\mathbb{Z}$.

(8) Show that the two multiplications on $\mathbb{H}$ in terms of $\mathbb{R}^4$ and $\mathbb{C}^2$ agree.

(9) Show that $S^3 \subset \mathbb{H}$ is not commutative. Determine its center.

(10) Show that the map explained in this chapter $\mu : S^3 \times S^3 \to S^3$ mapping (q_1, q_2) to $f(q_1) \cdot g(q_2)$ induces a map in homology mapping $[S^3, i_1]$ to $\deg f \cdot [S^3]$ and $[S^3, i_2]$ to $\deg g \cdot [S^3]$.

(11) Give a detailed proof of Proposition 11.4.

(12) Let $p : E \to M$ be a differentiable fibre bundle with E and M compact, M connected and fibre F. Show that
$$e(E) = e(M)e(F).$$
(Hint: You are allowed to use that M has a finite good atlas (for the definition see chapter 14).)

Chapter 12

Cohomology and Poincaré duality

Prerequisites: We assume that the reader knows what a smooth vector bundle is [**B-J**], [**Hi**].

1. Cohomology groups

In this chapter we consider another bordism group of stratifolds which at first glance looks like homology. It is only defined for smooth manifolds (without boundary). Similar groups were first introduced by Quillen [**Q**] and Dold [**D**]. They consider bordism classes of smooth manifolds instead of stratifolds.

The main difference between the new groups and homology is that we consider bordism classes of non-compact stratifolds. To obtain something non-trivial we require that the map $g : \mathbf{T} \to M$ is a proper map. We recall that a map between paracompact spaces is **proper** if the preimage of each compact space is compact. A second difference is that we only consider smooth maps. For simplicity we only define these bordism groups for oriented manifolds. (Each manifold is canonically homotopy equivalent to an oriented manifold, namely the total space of the tangent bundle, so that one can extend the definition to non-oriented manifolds using this trick, see the exercises in chapter 13.)

Definition: *Let M be an oriented smooth m-dimensional manifold without boundary. Then we define the* **integral cohomology group** $SH^k(M)$ *as* the group of bordism classes of proper smooth maps $g : \mathbf{S} \to M$, where $\mathbf{S}$ is an oriented regular stratifold of dimension $m - k$, addition is by disjoint union of maps and the inverse of $[\mathbf{S}, g]$ is $[-\mathbf{S}, g]$ (of course we also require that the maps for bordisms are proper and smooth and that the stratifolds are oriented and regular).

The reader might wonder why we required that M be oriented. The definition seems to work without this condition. This will become clear when we define induced maps. Then we will understand the relationship between $SH^k(M)$ and $SH^k(-M)$ better.

The relation between the grade, k, of $SH^k(M)$ and the dimension $m-k$ of representatives of the bordism classes looks strange but we will see that it is natural for various reasons.

If M is a point then $g : \mathbf{S} \to \text{pt}$ is proper if and only if $\mathbf{S}$ is compact. Thus
$$SH^k(\text{pt}) = SH_{-k}(\text{pt}) \cong \mathbb{Z}, \text{ if } k = 0, \text{ and } 0 \text{ if } k \neq 0.$$

In order to develop an initial feeling for cohomology classes, we consider the following situation. Let $p : E \to N$ be a k-dimensional, smooth, oriented vector bundle over an n-dimensional oriented smooth manifold. Then the total space E is a smooth $(k+n)$-dimensional manifold. The orientations of M and E induce an orientation on this manifold. The 0-section $s : N \to E$ is a proper map since $s(N)$ is a closed subspace. Thus
$$[N, s] \in SH^k(E)$$
is a cohomology class. This is the most important example we have in mind and will play an essential role when we define characteristic classes. A special case is given by a 0-dimensional vector bundle where $E = N$ and $p = \text{id}$. Thus we have for each smooth oriented manifold N the class $[N, \text{id}] \in SH^0(N)$, which we call $1 \in SH^0(N)$. Later we will define a multiplication on the cohomology groups and it will turn out that multiplication with $[N, \text{id}]$ is the identity, justifying the notation.

Is the class $[N, s]$ non-trivial? We will see that it is often non-trivial but it is zero if E admits a nowhere vanishing section $v : N \to E$. Namely then we obtain a zero bordism by taking the smooth manifold $N \times [0, \infty)$ and the map $G : N \times [0, \infty) \to E$ mapping $(x, t) \mapsto tv(x)$. The fact that v is nowhere vanishing implies that G is a proper map. Thus we have shown:

Proposition 12.1. *Let $p : E \to N$ be a smooth, oriented k-dimensional vector bundle over a smooth oriented manifold N. If E has a nowhere vanishing section v then $[N, s] \in SH^k(E)$ vanishes.*

In particular, if $[N, s]$ is non-trivial, then E does not admit a nowhere vanishing section.

In the following considerations and constructions it will be helpful for the reader to look at the cohomology class $[N, s] \in SH^k(E)$ and test the situation with this class.

2. Poincaré duality

Cohomology groups are, as indicated for example in Proposition 12.1, a useful tool. To apply this tool one has to find methods for their computation. We will do this in two completely different ways. The fact that they are so different is very useful since one can combine the information to obtain very surprising results like the vanishing of the Euler characteristic of odd-dimensional, compact, smooth manifolds.

The first tool, the famous Poincaré duality isomorphism, only works for compact, oriented manifolds and relates their cohomology groups to the homology groups. Whereas in the classical approach to (co)homology the duality theorem is difficult to prove, it is almost trivial in our context. The second tool is the Kronecker pairing which relates the cohomology groups to the dual space of the homology groups. This will be explained in chapter 14.

Let M be a compact oriented smooth m-dimensional manifold. (Here we recall that if we use the term manifold, then it is automatically without boundary; in this book, manifolds with boundary are always called c-manifolds. Thus a compact manifold is what in the literature is often called a **closed manifold**, a compact manifold without boundary.) If M is compact and $g : \mathbf{S} \to M$ is a proper map, then $\mathbf{S}$ is actually compact. Thus we obtain a homomorphism

$$P : SH^k(M) \to SH_{m-k}(M)$$

which assigns to $[\mathbf{S}, g] \in SH^k(M)$ the class $[\mathbf{S}, g]$ considered as element of $SH_{m-k}(M)$. Here we only forget that the map g is smooth and consider it as a continuous map.

Theorem 12.2. (Poincaré duality) *For a* **closed** *smooth oriented m-dimensional manifold M the map*

$$P : SH^k(M) \to SH_{m-k}(M)$$

is an isomorphism

Proof: For the proof we apply the following useful approximation result for continuous maps from a stratifold to a smooth manifold. It is another nice application of partitions of unity.

Proposition 12.3. *Let* $f : \mathbf{S} \to N$ *be a continuous map, which is smooth in an open neighbourhood of a closed subset* $A \subset \mathbf{S}$. *Then there is a smooth map* $g : \mathbf{S} \to N$ *which agrees with* f *on* A *and which is homotopic to* f *rel.* A.

Proof: The proof is the same as for a map f from a smooth manifold M to N in [**B-J**, Theorem 14.8]. More precisely, there it is proved that if we embed N as a closed subspace into an Euclidean space $\mathbb{R}^n$ then we can find a smooth map g arbitrarily close to f. The proof only uses that M supports a smooth partition of unity. Finally, sufficiently close maps are homotopic by ([**B-J**] Satz 12.9).
q.e.d.

As a consequence we obtain a similar result for *c*-stratifolds.

Proposition 12.4. *Let* $f : \mathbf{T} \to M$ *be a continuous map from a smooth c-stratifold* $\mathbf{T}$ *to a smooth manifold* M, *whose restriction to* $\partial \mathbf{T}$ *is a smooth map. Then* f *is homotopic rel. boundary to a smooth map.*

The proof follows from 12.3 using an appropriate closed subset in the collar of $\overset{\circ}{\mathbf{T}}$ for the subset A.

We apply this result to finish our proof. If $g : \mathbf{S} \to M$ represents an element of $SH_{m-k}(M)$, we can apply Proposition 12.3 to replace g by a homotopic smooth map g' and so $[\mathbf{S}, g] = P([\mathbf{S}, g'])$. This gives surjectivity of P. Similarly we use the relative version 12.4 to show injectivity. Namely, if for $[\mathbf{S}_1, g_1]$ and $[\mathbf{S}_2, g_2]$ in $SH^k(M)$ we have $P([\mathbf{S}_1, g_1]) = P([\mathbf{S}_2, g_2])$, there is a bordism $(\mathbf{T}, G)$ between these two pairs, where G is a continuous map whose restriction to the boundary is smooth. We apply Proposition 12.4 to replace G by a smooth map G' which agrees with the restriction of G on the boundary. Thus $[\mathbf{S}_1, g_1] = [\mathbf{S}_2, g_2] \in SH^k(M)$ and P is injective.
q.e.d.

By considering bordism classes of proper maps on $\mathbb{Z}/2$-oriented regular stratifolds we can define $\mathbb{Z}/2$-**cohomology groups** for arbitrary (non-oriented) smooth manifolds as we did in the integral case. The only difference is that we replace oriented regular stratifolds by $\mathbb{Z}/2$-oriented regular stratifolds which means that $\mathbf{S}^{n-1} = \varnothing$ and that no condition is placed on the orientability of the top stratum. The corresponding cohomology groups are denoted by
$$SH^k(M;\mathbb{Z}/2).$$

The proof of Poincaré duality works the same way for $\mathbb{Z}/2$-(co)homology:

Theorem 12.5. (Poincaré duality for $\mathbb{Z}/2$-(co)homology) *For a closed smooth oriented manifold M the map*
$$P: SH^k(M;\mathbb{Z}/2) \to SH_{m-k}(M;\mathbb{Z}/2)$$
is an isomorphism.

As mentioned above, we want to provide other methods for computing the cohomology groups. They are based on the same ideas as used for computing homology groups, namely to show that the cohomology groups fulfill axioms similar to the axioms of homology groups. One of the applications of these axioms will be an isomorphism between $SH^k(M) \otimes \mathbb{Q}$ and $\text{Hom}(SH_k(M), \mathbb{Q})$ and an isomorphism of $\mathbb{Z}/2$-vector spaces between $SH^k(M;\mathbb{Z}/2)$ and $\text{Hom}(SH_k(M;\mathbb{Z}/2), \mathbb{Z}/2)$. The occurrence of the dual spaces $\text{Hom}(SH_k(M), \mathbb{Q})$ and $\text{Hom}(SH_k(M;\mathbb{Z}/2), \mathbb{Z}/2)$ indicates a difference between the fundamental properties of homology and cohomology. The induced maps occurring should reverse their directions. We will see that this is the case.

3. The Mayer-Vietoris sequence

One of the most powerful tools for computing cohomology groups is, as it is for homology, the Mayer-Vietoris sequence. To formulate it we have to define for an open subset U of a smooth oriented manifold M the map induced by the inclusion $i: U \to M$. We equip U with the orientation induced from M. If $g: \mathbf{S} \to M$ is a smooth proper map we consider the open subset $g^{-1}(U) \subset \mathbf{S}$ and restrict g to this open subset. It is again a proper map (why?) and thus we define
$$i^*[\mathbf{S}, g] := [g^{-1}(U), g|_{g^{-1}(U)}].$$

This is obviously well defined and gives a homomorphism $i^* : SH^k(M) \to SH^k(U)$. This map reverses the direction of the arrows, as was motivated above. If V is an open subset of U and $j : V \to U$ is the inclusion, then by construction

$$j^* i^* = (ij)^*.$$

The next ingredient for the formulation of the Mayer-Vietoris sequence is the coboundary operator. We consider open subsets U and V in a smooth oriented manifold M, denote $U \cup V$ by X and define the coboundary operator

$$\delta : SH^k(U \cap V) \to SH^{k+1}(U \cup V)$$

as follows. We introduce the disjoint closed subsets $A := X - V$ and $B := X - U$. We choose a smooth map $\rho : U \cup V \to \mathbb{R}$ mapping A to 1 and B to -1. Now we consider $[\mathbf{S}, f] \in SH^k(U \cap V)$. Let $s \in (-1, 1)$ be a regular value of ρf. The preimage $\mathbf{D} := (\rho f)^{-1}(s)$ is an oriented regular stratifold of dimension $n-1$ sitting in $\mathbf{S}$. We define $\delta([\mathbf{S}, f]) := [\mathbf{D}, f|_{\mathbf{D}}] \in SH^{k+1}(X)$. It is easy to check that $f|_{\mathbf{D}}$ is proper.

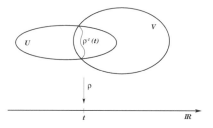

As with the definition of the boundary map for the Mayer-Vietoris sequence in homology, one shows that δ is well defined and that one obtains an exact sequence. For details we refer to Appendix B.

At first glance this definition of the coboundary operator looks strange since $f(\mathbf{D})$ is contained in $U \cap V$. But considered as a class in the cohomology of $U \cap V$ it is trivial. It is even zero in $SH^{k+1}(U)$ as well as in $SH^{k+1}(V)$. The reason is that in the construction of δ we can decompose $\mathbf{S}$ as $\mathbf{S}_+ \cup_\mathbf{D} \mathbf{S}_-$ with $\rho(\mathbf{S}_+) \geq s$ and $\rho(\mathbf{S}_-) \leq s$ (as for the boundary operator in homology we can assume up to bordism that there is a bicollar along $\mathbf{D}$). Then $(\mathbf{S}_-, f|_{\mathbf{S}_-})$ is a zero bordism of $(\mathbf{D}, f|_{\mathbf{D}})$ in U (note that $f|_{\mathbf{S}_-}$ is proper as a map into U and not into V). Similarly $(\mathbf{S}_+, f|_{\mathbf{S}_+})$ is a zero bordism of $(\mathbf{D}, f|_{\mathbf{D}})$ in V. But in $SH^{k+1}(U \cup V)$ it is in general non-trivial.

We summarize:

Theorem 12.6. (Mayer-Vietoris sequence for integral cohomology)
The following sequence is exact and commutes with induced maps:
$$\cdots \to SH^n(U \cup V) \to SH^n(U) \oplus SH^n(V)$$
$$\to SH^n(U \cap V) \xrightarrow{\delta} SH^{n+1}(U \cup V) \to \cdots .$$
The map $SH^n(U \cup V) \to SH^n(U) \oplus SH^n(V)$ is given by $\alpha \mapsto (j_U^*(\alpha), j_V^*(\alpha))$, the map from $SH^n(U) \oplus SH^n(V)$ to $SH^n(U \cap V)$ by $(\alpha, \beta) \mapsto i_U^*(\alpha) - i_V^*(\beta)$.

4. Exercises

(1) Compute the cohomology groups $SH^k(\mathbb{R}^n)$ for $k \geq 0$. (Hint: For $SH^0(\mathbb{R}^n)$ construct a map to $\mathbb{Z}$ by counting points with orientation in the preimage of a regular value. For degree > 0 apply Sard's theorem.) What happens for $k < 0$?

(2) Let $f : M \to N$ be a submersion (i.e., the differential df_x at each point $x \in M$ is surjective). Let $[g : \mathbf{S} \to N]$ be a cohomology class in $SH^k(N)$. Show that the pull-back $\{(x, y) \in (M \times \mathbf{S}) \mid (f(x) = g(y)\}$ is a stratifold and that the projection to the first factor is a proper map. Show that this construction gives a well defined homomorphism $f^* : SH^k(N) \to SH^k(M)$. (This is a special case of the induced map which we will define later.)

(3) Let M be a smooth manifold. Show that the map $p^* : SH^k(M) \to SH^k(M \times \mathbb{R})$ is injective. (Hint: Construct a map $SH^k(M \times \mathbb{R}) \to SH^k(M)$ by considering for $[g : \mathbf{S} \to M \times \mathbb{R}]$ a regular value of $p_2 g$.) We will see later that p^* is an isomorphism; try to prove this directly.

Chapter 13

Induced maps and the cohomology axioms

Prerequisites: in this chapter we apply one of the most powerful tools from differential topology, namely transversality. The necessary information can be found in [**B-J**], [**Hi**].

1. Transversality for stratifolds

We recall the basic definitions and results concerning transversality of manifolds. Let M, P and Q be smooth manifolds of dimensions m, p and q, and let $f : P \to M$ and $g : Q \to M$ be smooth maps. Then we say that f is transverse to g if for all $x \in P$ and $y \in Q$ with $f(x) = g(y) = z$ we have $df(T_xP) + dg(T_yQ) = T_zM$. If $g : Q \to M$ is the inclusion of a point z in M then this condition means that z is a regular value of f. It is useful to note that the transversality condition is equivalent to the property that $f \times g : P \times Q \to M \times M$ is transverse to the diagonal $\Delta = \{(x,x)\} \subset M \times M$. Similarly, as for preimages of regular values, one proves that the **pull-back**, which we denote here by $(P, f) \pitchfork (Q, g) := \{(x, y) \in (P \times Q) \mid f(x) = g(y)\}$, is a smooth submanifold of $P \times Q$ of dimension $p+q-m$ [**B-J**], [**Hi**]. We call $(P, f) \pitchfork (Q, g)$ the **transverse intersection** of (P, f) and (Q, g). Later on we will generalize this construction to the case, where P is a stratifold.

If all three manifolds are oriented then there is a canonical orientation on $(P, f) \pitchfork (Q, g)$. To define this we begin with the case where g is an embedding. In this case we consider the normal bundle of $g(Q)$ and orient it in such a way that the concatenation of the orientations of TQ and the normal bundle give the orientation of M. In turn if an orientation of the normal bundle and of M is given we obtain an induced compatible orientation of Q. Now we note that in this case $(P, f) \pitchfork (Q, g)$ is diffeomorphic to a submanifold of P under the projection to the first factor and the normal bundle of $(P, f) \pitchfork (Q, g)$ in P is the pull-back of the normal bundle of Q in M and we equip it with the induced orientation, i.e., the orientation such that the isomorphism between the fibres of the normal bundle induced by the differential of f is orientation-preserving. By the considerations above this orientation of the normal bundle induces an orientation on $(P, f) \pitchfork (Q, g)$. If g is not an embedding we choose an embedding of Q into $\mathbb{R}^N$ (equipped with the canonical orientation) for some integer N and thicken M and P, replacing them by $M \times \mathbb{R}^N$ and $P \times \mathbb{R}^N$, and replace f by $f \times \mathrm{id}$. The map given by g on the first component and by the embedding to $\mathbb{R}^N$ on the second gives an embedding of Q into $M \times \mathbb{R}^N$ and $f \times \mathrm{id}$ is transverse to this embedding and the preimage is canonically diffeomorphic to $(P, f) \pitchfork (Q, g)$. Thus the construction above gives an orientation on $(P, f) \pitchfork (Q, g)$. This orientation depends neither on the choice of N nor the embedding to $\mathbb{R}^N$, since any two such embeddings are isotopic if we make N large enough by stabilization passing from $\mathbb{R}^N$ to $\mathbb{R}^{N+1}$. This definition of an induced orientation has the useful property that if $f' : P' \to P$ is another smooth map transverse to $(P, f) \pitchfork (Q, g)$, then the induced orientations on $(P', f') \pitchfork ((P', f) \pitchfork (Q, g))$ and $(P', ff') \pitchfork (Q, g)$ agree.

To shorten notation we often write $f \pitchfork g$ instead of $(P, f) \pitchfork (Q, g)$. If M, P and Q are oriented we mean this manifold with the induced orientation.

If we replace P by a smooth c-manifold with boundary and f is a smooth map transverse to g and also $f|_{\partial P}$ is transverse to g, then $(P, f) \pitchfork (Q, g) := \{(x, y) \in (P \times Q) \mid f(x) = g(y)\}$ is a smooth c-manifold of dimension $p+q-m$ with boundary $f|_{\partial P} \pitchfork g$. We obtain a similar statement if instead of admitting a boundary for P we replace Q by a c-manifold with boundary and require that f is transverse to the smooth c-map g as well as being transverse to $g|_{\partial Q}$.

The transversality theorem states that if $f : P \longrightarrow M$ and $g : Q \longrightarrow M$ are smooth maps then f is homotopic to f' such that f' is transverse to g [**B-J**], [**Hi**]. More generally, if $A \subset P$ is a closed subset and for some open neighbourhood U of A the maps $f|_U$ and g are transverse, then f is homotopic rel. A (i.e., the homotopy maps $(x, t) \in A \times I$ to $f(x)$) to f' such that f' is transverse to g.

A similar argument implies the following statement:

Theorem 13.1. *Let $f : P \longrightarrow M$ and $g_1 : Q_1 \longrightarrow M, \ldots, g_r : Q_r \longrightarrow M$ be smooth maps such that for some closed subset $\Lambda \subset P$ and open neighbourhood U of Λ the maps $f|_U$ and g_i are transverse for $i = 1, \ldots, r$. Then f is homotopic to f' rel. Λ in such a way that f' is transverse to g_i for all i.*

We want to generalize this argument to maps $f : P \longrightarrow M$, where as before P is a smooth manifold, and $g : \mathbf{S} \longrightarrow M$ is a morphism from a stratifold to M. The definition for transversality above generalizes to this situation. Equivalently we say that f is **transverse** to g if and only if f is transverse to restrictions of g to all strata. If f is transverse to g, then we obtain a stratifold denoted by $g \pitchfork f$ whose underlying space is $\{(x,y) \in \mathbf{S} \times P \mid f(x) = g(y)\}$. The algebra is given by $\mathbf{C}(g \pitchfork f)$, the restriction of the functions in $\mathbf{S} \times P$ to this space. The argument for showing that this is a stratifold is the same as for the special case of the preimage of a regular value (Proposition 2.5). The strata of $g \pitchfork f$ are $g|_{\mathbf{S}^i} \pitchfork f$. The dimension of $g \pitchfork f$ is $\dim P + \dim \mathbf{S} - \dim M$. If $\mathbf{S}$ is a regular stratifold, then $g \pitchfork f$ is regular, the isomorphisms of appropriate local neighborhoods of the strata with a product being given by restrictions of the corresponding isomorphisms for $\mathbf{S} \times P$.

As a consequence of the transversality theorem for manifolds we see:

Theorem 13.2. *Let $f : P \longrightarrow M$ be a smooth map from a smooth manifold P to M and $g : \mathbf{S} \longrightarrow M$ be a morphism from a stratifold $\mathbf{S}$ to M. Let A be a closed subset of P and U an open neighbourhood such that $f|_U$ is transverse to g. Then f is homotopic rel. A to f' such that f' is transverse to g.*

Proof: We simply apply Theorem 13.1 to replace f by f' (homotopic to f rel. A) such that f' is transverse to all $g|_{\mathbf{S}^i}$.
q.e.d.

2. The induced maps

We return to our construction of cohomology and define the induced maps. Let $f : N \to M$ be a smooth map between oriented manifolds and let $[\mathbf{S}, g]$ be an element of $SH^k(M)$. Then we replace f by a homotopic map

$f' : N \to M$ which is transverse to g and consider $f' \pitchfork g$. This is a regular stratifold of dimension $n + \dim \mathbf{S} - m = n + m - k - m = n - k$. The stratum of dimension $n - k - 1$ is empty. The projection to N gives a map $g' : g \pitchfork f' \to N$. This is a proper map (why?). The orientations of M, N and $\mathbf{S}$ induce an orientation of $f' \pitchfork g$, as explained above. **This is the place where we use the orientation of the manifold** M. Thus the pair $(g \pitchfork f', g')$ represents an element of $SH^k(N)$.

Using Theorem 13.2 we see that the bordism class of $(g \pitchfork f', g')$ is unchanged if we choose another map f'_1 homotopic to f and transverse to g. Namely then f' and f'_1 are homotopic. We can assume that this homotopy is a smooth map, and that there is an $\epsilon > 0$ such that $h(x, t) = f'(x)$ for $t < \epsilon$ and $h(x, t) = f'_1(x)$ for $t > 1 - \epsilon$ (such a homotopy is often called a technical homotopy). By Theorem 13.2 we can further assume that this homotopy h is transverse to g. Then $(g \pitchfork h, g')$ is a bordism between $(g \pitchfork f', g')$ and $(g \pitchfork f'_1, g')$.

For later use (the proof of Proposition 13.5) we note that this argument implies that the induced map is a homotopy invariant.

Next we show that if $(\mathbf{S}_1, g_1)$ and $(\mathbf{S}_2, g_2)$ are bordant, then $(g_1 \pitchfork f', g'_1)$ is bordant to $(g_2 \pitchfork f', g'_2)$, where f' is homotopic to f and transverse to g_1 and to g_2 simultaneously (by the argument above we are free in the choice of the map which is transverse to a given bordism class). Let $(\mathbf{T}, G)$ be a bordism between $(\mathbf{S}_1, g_1)$ and $(\mathbf{S}_2, g_2)$. Then again using the fact that we are free in the choice of f' we assume that f' is also transverse to G. Then $(G \pitchfork f', G')$ is a bordism between $(g_1 \pitchfork f', g'_1)$ and $(g_2 \pitchfork f', g'_2)$. Thus we obtain a well defined induced map

$$f^* : SH^k(M) \to SH^k(N)$$

mapping

$$[\mathbf{S}, g] \mapsto [g \pitchfork f', g']$$

where f' is transverse to g and g' is the restriction of the projection onto N. This construction respects disjoint unions and so we have defined the **induced homomorphism in cohomology** for a smooth map $f : N \to M$. As announced above this induced map in cohomology reverses its direction. By construction this definition agrees for inclusions with the previous definition used in the formulation of the Mayer-Vietoris sequence. Here one has to be careful with the orientation and we suggest that the reader checks that the conventions lead to the same definition.

2. The induced maps

The role of the orientation of the manifolds is reflected by the following induced maps. Let $f : N \to M$ be an orientation-preserving diffeomorphism. Then by construction $f^*([\mathbf{S}, g]) = [\mathbf{S}, f^{-1}g]$. If f reverses the orientation, $f^*([\mathbf{S}, g]) = [-\mathbf{S}, f^{-1}g]$. **In particular if we consider as f the identity map from M to $-M$ equipped with opposite orientation, we see that $f^*([\mathbf{S}, g]) = [-\mathbf{S}, g]$.** (In this context one should not write id for the identity map, whose name in the oriented world should be reserved for the identity map from M to M, where both are equipped with the same orientation.)

If M and N are not oriented the same construction gives us an induced map
$$f^* : SH^k(M; \mathbb{Z}/2) \to SH^k(N; \mathbb{Z}/2)$$
mapping
$$[\mathbf{S}, g] \mapsto [g \pitchfork f', g'].$$

An important case of an induced map is the situation considered in the previous chapter of a smooth oriented vector bundle $p : E \to N$ of rank k over an oriented manifold. We introduced the cohomology class $[N, s] \in SH^k(E)$. We want to look at $s^*([N, s]) \in SH^k(N)$. To obtain this class we have to approximate s by another map s' which is transverse to $s(N) \subset E$ (one can actually find s' which is again a section [**B-J**]). Then $s'(N) \pitchfork s(N)$ is a smooth submanifold of $s(N) = N$ of dimension $n - k$. Let $i : s'(N) \pitchfork s(N) \to s(N) = N$ be the inclusion; then we obtain
$$s^*([N, s]) = [s'(N) \pitchfork s(N), i] \in SH^k(N).$$

This class is called the **Euler class** of E and is abbreviated as
$$e(E) := s^*([N, s]) = [s'(N) \pitchfork s(N), i] \in SH^k(N).$$

In the next chapter we will investigate this class in detail. From Proposition 12.1 we conclude:

Proposition 13.3. *Let $p : E \to N$ be a smooth oriented k-dimensional vector bundle. If E has a nowhere vanishing section then $e(E) = 0$.*

3. The cohomology axioms

We will now formulate and prove properties of cohomology groups which are analogous to the axioms of a homology theory. Apart from the fact that induced maps change direction the main difference is that we have only defined cohomology groups for smooth manifolds and induced maps of smooth maps.

If $f : N \to M$ and $h : P \to N$ are smooth maps, such that f is transverse to $g : \mathbf{S} \to M$, where $\mathbf{S}$ is a regular stratifold, and h is transverse to $g' : f \pitchfork g \to N$, then $fh : P \to M$ is transverse to $g : \mathbf{S} \to M$ and $fh \pitchfork g = h \pitchfork g'$ (with induced orientations as explained at the beginning of this chapter). This implies the following:

Proposition 13.4. *Let $f_1 : M_1 \to M_2$ and $f_2 : M_2 \to M_3$ be smooth maps. Then*
$$f_1^* f_2^* = (f_2 f_1)^*.$$

Furthermore by definition:
$$\mathrm{id}^* = \mathrm{id}.$$

Here we stress again that we have reserved the name id for the identity map from M to M, both equipped with the same orientation!

Apart from the change of direction, these are the properties of a functor assigning to a smooth manifold an abelian group and to a smooth map a homomorphism between these groups reversing its direction. To distinguish it from a functor in the previous sense we call it a **contravariant functor**. To make notation more symmetric, a functor in the previous sense is often also called a **covariant functor**.

To compare the Mayer-Vietoris sequence of different spaces it is useful to know that induced maps commute with the coboundary operator. Since the construction of the coboundary operator for cohomology is completely analogous to that for homology the same argument implies this statement.

The property of a contravariant functor (Proposition 13.4) is—in analogy to homology—the first fundamental property of a cohomology theory. The other two are the homotopy axiom and the Mayer-Vietoris sequence which we have already constructed. The homotopy axiom was also already shown when we proved that the induced map is well defined:

Proposition 13.5. *Let $f : N \to M$ and $g : N \to M$ be homotopic smooth maps. Then*
$$f^* = g^* : SH^k(M) \to SH^k(M).$$

A contravariant functor $SH^k(M)$ assigning to each smooth manifold abelian groups and to each smooth map an induced map such that the statements of Theorems 12.6, 13.4, 13.5 hold and where the coboundary operator in the Mayer-Vietoris sequence commutes with induced maps, is called a **cohomology theory** for smooth manifolds and smooth maps. Thus cohomology as defined here is a cohomology theory.

As for homology one can use the cohomology axioms to compute the cohomology groups for many spaces, such as spheres and complex projective spaces. For compact oriented manifolds without boundary one can use Poincaré duality and reduce it to the computation of homology groups.

4. Exercises

(1) Construct an orientation on the total space of the tangent bundle of a smooth manifold M, which has the property that for an open subset $U \subset M$ the construction agrees with the restriction of the orientation to the total space of the tangent bundle of U and that if $f : M \to N$ is a diffeomorphism, then $df : TM \to TN$ is an orientation-preserving diffeomorphism. We further require that for $M = \mathbb{R}^n$ the orientation of $T\mathbb{R}^n = \mathbb{R}^n \times \mathbb{R}^n$ agrees with the standard orientation. Show that this orientation is unique.

(2) If M is a non-orientable manifold define $SH^k(M)$ by $SH^k(TM)$. Show that if M is oriented then $p^* : SH^k(M) \to SH^k(TM)$, where TM is oriented as above, is an isomorphism. This way we extend the definition for oriented manifolds to arbitrary manifolds.

(3) For a map $f : M \to N$ consider the map $\hat{f} := sfp : TM \to TN$, where $p : TM \to M$ is the projection and $s : N \to TN$ is the zero-section. Define the induced map $f^* : SH^k(N) = SH^k(TN) \to SH_k(M) = SH^k(TM)$ as $\hat{f}^*$. Show that this way we obtain a contravariant functor for arbitrary smooth manifolds. Show that this is a cohomology theory which extends our definition for oriented manifolds. Could you use the differential df instead of $\hat{f}$?

(4) Compute the cohomology groups of $\mathbb{RP}^n$.

(5) Let $p_1 : M \times N \to M$ and $p_2 : M \times N \to N$ be the projections on the first and second factor. Show that for $[S, f] \in H^k(M)$ and

$[S', f'] \in H^r(N)$
$$p_1^*([S, f]) = [S \times N, f \times \text{id}]$$
and
$$p_2^*([S', f']) = [M \times S', \text{id} \times f'].$$

Chapter 14

Products in cohomology and the Kronecker pairing

1. The cross product and the Künneth theorem

So far the basic structure of cohomology is completely analogous to that of homology. The essential difference was the change of the direction of the maps induced between cohomology groups. In this chapter we will introduce a new structure on cohomology called the cup product. As before we assume in this chapter that all manifolds are oriented.

The cup product is derived from the $\times$-product which is defined up to sign as was the $\times$-product in homology. Let M and N be smooth oriented manifolds of dimension m and n respectively. If $[\mathbf{S}_1, g_1] \in SH^k(M)$ and $[\mathbf{S}_2, g_2] \in SH^\ell(N)$, we define

$$[\mathbf{S}_1, g_1] \times [\mathbf{S}_2, g_2] := (-1)^{\ell(m-k)}[\mathbf{S}_1 \times \mathbf{S}_2, g_1 \times g_2] \in SH^{k+\ell}(M \times N).$$

The sign looks strange at first glance, but it is needed to give a pleasant expression when interchanging the factors, as we will discuss in the next paragraph. As in homology the $\times$-**product** or **cross product**

$$\times : SH^k(M) \times SH^\ell(N) \to SH^{k+\ell}(M \times N)$$

is a **bilinear and associative** map (check associativity). It is also **natural**, i.e., for a smooth map $f : M' \to M$ and $g : N' \to N$ and $\alpha \in SH^k(M)$ and

$\beta \in SH^\ell(N)$ we have:
$$(f \times g)^*(\alpha \times \beta) = f^*(\alpha) \times g^*(\beta),$$
exercise (6).

In the same way one defines the ×-product for $\mathbb{Z}/2$-cohomology (one can of course omit the signs here):
$$\times : SH^k(M; \mathbb{Z}/2) \times SH^\ell(N; \mathbb{Z}/2) \to SH^{k+\ell}(M \times N; \mathbb{Z}/2).$$
This fulfills the same properties as the product above.

As announced, we study the behavior of the ×-product under a change of the factors. For this we consider the flip diffeomorphism $\tau : N \times M \to M \times N$ mapping (x, y) to (y, x), where M and N are oriented manifolds having dimensions m and n, respectively. Then τ changes the orientation by $(-1)^{mn}$. Thus by the interpretation of induced maps for diffeomorphisms, if $[\mathbf{S}, f] \in SH^k(M)$ and $[\mathbf{S}', f'] \in SH^\ell(N)$, then
$$\tau^*([\mathbf{S} \times \mathbf{S}', f \times f']) = (-1)^{mn}[\mathbf{S} \times \mathbf{S}', \tau^{-1}(f \times f')].$$
To compare this with $[\mathbf{S}' \times \mathbf{S}, f' \times f]$ we consider the flip map τ' from $\mathbf{S} \times \mathbf{S}'$ to $\mathbf{S}' \times \mathbf{S}$ and note that $\tau^{-1}(f \times f') = (f' \times f)\tau'$. Since τ' changes the orientation by the factor $(-1)^{\dim \mathbf{S} \dim \mathbf{S}'} = (-1)^{(m-k)(n-\ell)}$, we conclude that
$$\tau^*([\mathbf{S} \times \mathbf{S}', f \times f']) = (-1)^{mn}(-1)^{(m-k)(n-\ell)}[\mathbf{S}' \times \mathbf{S}, (f' \times f)]$$
$$= (-1)^{m\ell+nk+k\ell}[\mathbf{S}' \times \mathbf{S}, (f' \times f)].$$
Now we combine these signs with the sign occurring in the definition of the ×-product to obtain:
$$\tau^*([\mathbf{S}, f] \times [\mathbf{S}', f']) = \tau^*((-1)^{\ell(m-k)}([\mathbf{S} \times \mathbf{S}', f \times f'])) = (-1)^{nk}[\mathbf{S}' \times \mathbf{S}, f' \times f]$$
$$= (-1)^{k\ell}[\mathbf{S}', f'] \times [\mathbf{S}, f].$$
Thus we have the equality
$$\tau^*([\mathbf{S}, f] \times [\mathbf{S}', f']) = (-1)^{k\ell}([\mathbf{S}', f'] \times [\mathbf{S}, f]).$$

The ×-product is a very useful tool. For example – as for homology – the ×-product is used in a Künneth theorem for rational cohomology and for $\mathbb{Z}/2$-cohomology. Here we define the **rational cohomology** groups $SH^k(M; \mathbb{Q}) := SH^k(M) \otimes \mathbb{Q}$. By elementary algebraic considerations similar to the arguments for rational homology groups one shows that rational cohomology fulfills the axioms of a cohomology theory. The proof of the Künneth Theorem would be the same as for homology if we had a comparison theorem like Corollary 9.4. The proof of Corollary 9.4 used the fact that homology groups are compactly supported. This is not the case for cohomology groups. But the inductive proof of Corollary 9.4 based on the 5-Lemma goes through in cohomology if we can cover M by finitely many open subsets U_i such that we know that the natural transformation is an isomorphism for all finite intersections of these subsets. This leads to the concept of a **good atlas** of a smooth manifold M. This is an atlas $\{\varphi_i : U_i \to V_i\}$ such that all non-empty finite intersections of the U_i are diffeomorphic to $\mathbb{R}^m$. But $\mathbb{R}^m$ is homotopy equivalent to a point and, if we assume that for a point we have an isomorphism between the cohomology theories, the induction argument for the proof of Corollary 9.4 works for cohomology, if M has a finite good atlas:

Proposition 14.1. *Let M be a smooth oriented manifold admitting a finite good atlas. Let h and h' be cohomology theories and $\tau : h \to h'$ be a natural transformation which for a point is an isomorphism in all degrees. Then $\tau : h^k(M) \to (h')^k(M)$ is an isomorphism for all k.*

One can show that all smooth manifolds admit a good atlas (compare [**B-T**, Theorem 5.1]). In particular all compact manifolds admit a finite good atlas.

If we combine Proposition 14.1 with the argument for the Künneth isomorphism in homology we obtain:

Theorem 14.2. (Künneth Theorem for cohomology) *Let M be a smooth oriented manifold admitting a finite good atlas. Then for F equal to $\mathbb{Z}/2$ or equal to $\mathbb{Q}$, for each smooth oriented manifold N the $\times$-product induces an isomorphism*

$$\times : \bigoplus_{i+j=k} SH^i(M; F) \otimes_F SH^j(N; F) \to SH^k(M \times N; F).$$

If all cohomology groups of N are torsion-free and finitely generated, then the same holds for integral cohomology.

2. The cup product

The following construction with the cross product is the main difference between homology and cohomology since it can only be carried out for cohomology. Let $\Delta : M \to M \times M$ be the diagonal map $x \mapsto (x, x)$. Then we define the **cup product** as follows

$$\smile : SH^k(M) \times SH^\ell(M) \to SH^{k+\ell}(M)$$

$$([\mathbf{S}_1, g_1], [\mathbf{S}_2, g_2]) \mapsto \Delta^*([\mathbf{S}_1, g_1] \times [\mathbf{S}_2, g_2]).$$

The cup product has the following property, which one often calls **graded commutativity**:

$$[\mathbf{S}_1, g_1] \smile [\mathbf{S}_2, g_2] = (-1)^{k\ell}[\mathbf{S}_2, g_2] \smile [\mathbf{S}_1, g_1].$$

This follows from the behavior of the $\times$-product under the flip map τ shown above together with the fact that $\tau\Delta = \Delta$.

There is also a neutral element, namely the cohomology class $[M, \mathrm{id}] \in SH^0(M)$. To see this we consider $[\mathbf{S}, g] \in SH^k(M)$. Then $[M, \mathrm{id}] \times [\mathbf{S}, g] = [M \times \mathbf{S}, \mathrm{id} \times g]$. To determine $\Delta^*([M \times \mathbf{S}, \mathrm{id} \times g])$ we note that $\mathrm{id} \times g$ is transverse to Δ and so $\Delta^*([M \times \mathbf{S}, \mathrm{id} \times g]) = [\mathbf{S}, g]$, i.e., $[M, \mathrm{id}]$ is a neutral element. This property justifies our previous notation:

$$1 := [M, \mathrm{id}] \in SH^0(M)$$

and we have

$$1 \smile [\mathbf{S}, g] = [\mathbf{S}, g].$$

Similarly, one shows

$$[\mathbf{S}, g] \smile 1 = [\mathbf{S}, g].$$

Furthermore we note that the naturality of the $\times$-product implies the naturality of the $\times$-product:

$$f^*([\mathbf{S}_1, g_1] \smile [\mathbf{S}_2, g_2]) = f^*([\mathbf{S}_1, g_1]) \smile f^*([\mathbf{S}_2, g_2]).$$

From the corresponding properties of the $\times$-product one concludes that the cup product is bilinear and associative.

We defined the cup product in terms of the $\times$-product. One can also derive the $\times$-product from the cup product. Let $\alpha \in SH^k(M)$ and $\beta \in H^\ell(N)$, then

$$p_1^*(\alpha) \cup p_2^*(\beta) = \alpha \times \beta.$$

This is exercise (5).

The following is a useful observation for the computation of the $\cup$-product. Let M be an oriented manifold and suppose that $[N_1, g_1] \in SH^k(M)$ and $[N_2, g_2] \in SH^\ell(M)$ are cohomology classes with N_i smooth manifolds. Then we can obtain the cup product by considering as before $g := g_1 \times g_2$. But instead of making the diagonal transverse to g and then

2. The cup product

taking the transverse intersection we can keep the diagonal Δ unchanged, approximate g instead by a map g' transverse to Δ and take the transverse intersection. It is easy to use the transversality theorem to prove the existence of a bordism between the two cohomology classes obtained by making Δ transverse to g or by making g transverse to Δ. Furthermore we can interpret the latter transverse intersection as the transverse intersection of g_1 and g_2, i.e., we approximate g_1 by g_1' transverse to g_2 and then we obtain:

Lemma 14.3. *Let $[N_1, g_1] \in SH^k(M)$ and $[N_2, g_2] \in SH^\ell(M)$ be cohomology classes with N_i smooth manifolds such that g_1 is transverse to g_2. Then*
$$[N_1, g_1] \smile [N_2, g_2] = [g_1 \pitchfork g_2, g_1 p_1],$$
where p_1 is the projection to the first factor.

A priori this identity is only clear up to sign and we have to show that the sign is $+$. To do this, it is enough to consider the case, where g_1 and g_2 are embeddings (replace M by $M \times \mathbb{R}^N$ for some large N and approximate g_i by embeddings) and after identifying the N_i with their images under g_i, we assume that the N_i are submanifolds of M. The orientation of $N_1 \cap N_2 \subset N_1$ (which with the inclusion to M represents $[g_1 \pitchfork g_2, g_1 p_1]$) at $x \in N_1 \cap N_2$ is given by requiring that $T_x N_1 = T_x(N_1 \cap N_2) \oplus \nu_x(N_2, M)$ (where $\nu(N_2, M)$ is the normal bundle of N_2 in M) preserves the orientations induced from the orientation of N_i and M. On the other hand $\Delta^*([N_1 \times N_2, g_1 \times g_2])$ is represented by $N_1 \cap N_2$ together with the inclusion to M which we identify with $\Delta(M)$. The orientation at $x \in N_1 \cap N_2$ of $N_1 \cap N_2 \subset M$ is given by requiring that the decomposition $T_x(N_1 \cap N_2) \oplus \nu_x(N_1 \times N_2, M \times M) = T_x \Delta(M) = T_x M$ preserves the orientation. We have to determine the orientation of $\nu_x(N_1 \times N_2, M \times M)$ in terms of the orientations of $\nu_x(N_1, M)$ and $\nu_x(N_2, M)$. Comparing the orientations of $T_x N_1 \oplus \nu_x(N_1, M) \oplus T_x N_2 \oplus \nu_x(N_2, M) = T_x M \oplus T_x M$ and $T_x N_1 \oplus T_x N_2 \oplus \nu_x(N_1, M) \oplus \nu_x(N_2, M) = T_{(x,x)}(M \times M)$, we see that as oriented vector spaces $\nu_x(N_1 \times N_2, M \times M) = (-1)^{(m-n_1)n_2} \nu_x(N_1, M) \oplus \nu_x(N_2, M)$, where $m = \dim M$. Combining this with the identity
$$T_x(N_1 \cap N_2) \oplus \nu_x(N_1, M) \oplus \nu_x(N_2, M) = T_x M$$
we obtain
$$(-1)^{(m-n_1)n_2} T_x(N_1 \cap N_2) \oplus \nu_x(N_1, M) \oplus \nu_x(N_2, M) = T_x M = T_x N_1 \oplus \nu_x(N_1, M).$$
Comparing this with the orientation of $N_1 \cap N_2 \subset N_1$ we conclude that
$$(-1)^{(m-n_1)n_2} T_x(N_1 \cap N_2) \oplus \nu_x(N_1, M) \oplus \nu_x(N_2, M)$$
$$= T_x(N_1 \cap N_2) \oplus \nu_x(N_2, M) \oplus \nu_x(N_1, M)$$

and so we conclude that the orientations differ by
$$(-1)^{(m-n_1)n_2}(-1)^{(m-n_1)(m-n_2)} = (-1)^{m(m-n_1)} = (-1)^{mk},$$
where $k = m - n_1$. This is the sign we introduced when defining the ×-product and so we have shown that the sign in the formula is correct.

For example we can use this to compute the cup product structure for the complex projective spaces $\mathbb{CP}^n$. Since these are closed oriented smooth manifolds we have by Poincaré duality $SH^{2k}(\mathbb{CP}^n) = SH_{2n-2k}(\mathbb{CP}^n)$, which by Theorem 8.8 is $\mathbb{Z}$ generated by $[\mathbb{CP}^{n-k}, i]$, where i is the inclusion mapping $[z_0, \ldots, z_{n-k}]$ to $[z_0, \ldots, z_{n-k}, 0, \ldots, 0]$. To compute the cup product $[\mathbb{CP}^{n-k}, i] \smile [\mathbb{CP}^{n-l}, j]$ we have to replace i by a map which is transverse to j. This can easily be done by choosing an appropriate alternative embedding, namely $i'([z_0, \ldots, z_{n-k}]) := [0, \ldots, 0, z_0, \ldots, z_{n-k}]$. This represents the same homology class since the inclusions are homotopic. The map i' is transverse to j and so the cup product is represented by $[i'(\mathbb{CP}^{n-k}) \cap j(\mathbb{CP}^{n-l}), s]$, where s is again the inclusion. The intersection is $\mathbb{CP}^{n-k-l}$ and the map is up to a permutation the standard embedding. We conclude that
$$[\mathbb{CP}^{n-k}, i] \smile [\mathbb{CP}^{n-l}, i] = [\mathbb{CP}^{n-k-l}, i].$$

As a consequence we put $x := [\mathbb{CP}^{n-1}, i] \in SH^2(\mathbb{CP}^n)$ and conclude:
$$x^r = [\mathbb{CP}^{n-r}, i] \in SH^{2r}(\mathbb{CP}^n),$$
where x^r stands for the r-fold cup product. In particular $x^n = [\mathbb{CP}^0, i]$, the canonical generator of $SH^{2n}(\mathbb{CP}^n)$.

It is useful to collect all cohomology groups into a direct sum and denote it by
$$SH^*(M) := \bigoplus_k SH^k(M).$$
The cup product induces a ring structure on $SH^*(M)$ by:
$$(\sum_i \alpha_i)(\sum_j \beta_j) := \sum_k (\sum_{i+j=k} \alpha_i \smile \beta_j),$$
where $\alpha_i \in SH^i(M)$ and $\beta_j \in SH^j(M)$. In this way we consider $SH^*(M)$ a ring called the **cohomology ring**. The computation above for the complex projective spaces can be reformulated as:
$$SH^*(\mathbb{CP}^n) = \mathbb{Z}[x]/x^{n+1}.$$
This ring is called a truncated polynomial ring.

We also introduce the $\mathbb{Z}/2$-cohomology ring as
$$SH^*(M; \mathbb{Z}/2) := \bigoplus_k SH^k(M; \mathbb{Z}/2).$$

Using a similar argument one shows that
$$SH^*(\mathbb{RP}^n; \mathbb{Z}/2) = (\mathbb{Z}/2)[x]/x^{n+1},$$
where $x \in SH^1(\mathbb{RP}^n; \mathbb{Z}/2)$ is the non-trivial element.

3. The Kronecker pairing

Now we can prove the announced relation between cohomology and homology groups. Let M be an oriented smooth m-dimensional manifold. The first step is the construction of the so-called Kronecker homomorphism from $SH^k(M)$ to $\mathrm{Hom}\,(SH_k(M), \mathbb{Z})$. The map is induced by a bilinear map $SH^k(M) \times SH_k(M) \to \mathbb{Z}$. To describe this let $[\mathbf{S}_1, g_1] \in SH^k(M)$ be a cohomology class and $[\mathbf{S}_2, g_2] \in SH_k(M)$ be a homology class. Applying Proposition 12.4 we can approximate g_2 by a smooth map and so we assume from now on that g_2 is smooth. We consider $g = g_1 \times g_2 : (-1)^{mk} \mathbf{S}_1 \times \mathbf{S}_2 \to M \times M$. The sign changing the orientation of $\mathbf{S}_1 \times \mathbf{S}_2$ is compatible with the sign introduced in the definition of the $\times$-product.

Let $\Delta : M \to M \times M$ be the diagonal map. We want to approximate Δ by a smooth map Δ' which is transverse to $g_1 \times g_2$ in such a way that the transverse intersection $\Delta' \pitchfork (g_1 \times g_2)$ is compact. To achieve this we note that since g_1 is proper, and $\mathbf{S}_2$ is compact the intersection $\mathrm{im}(g_1 \times g_2) \cap \mathrm{im}(\Delta)$ is compact. Namely, we define $C_0 := \{x \in \mathbf{S}_1 \mid g_1(x) \in \mathrm{im}(g_2)\}$, which is compact since $\mathbf{S}_2$ is compact and g_1 is proper. Thus $g_1 \times g_2(C_0 \times \mathbf{S}_2)$ is compact. But $\mathrm{im}(g_1 \times g_2) \cap \mathrm{im}(\Delta)$ is a closed subset of $(g_1 \times g_2)(C_0 \times \mathbf{S}_2)$ and so is compact. Since Δ is proper, $C_1 := \Delta^{-1}(\mathrm{im}(g_1 \times g_2) \cap \mathrm{im}(\Delta))$ is compact. We choose compact subsets $C_2 \subset C_3 \subset M$ such that $C_1 \subset \mathring{C}_2$ and $C_2 \subseteq \mathring{C}_3$. Then $A := M - \mathring{C}_2$ is a closed subset which is contained in the open subset $U := M - C_1$. Since $\mathrm{im}(g_1 \times g_2) \cap \Delta(U) = \emptyset$, the map $\Delta|_U$ is transverse to $g_1 \times g_2$. We approximate Δ by a transverse map Δ', which agrees with A on Δ. By construction, $\Delta' \pitchfork (g_1 \times g_2) \subset C_2 \times \mathbf{S}_1 \times \mathbf{S}_2$. The set $D := \{x \in \mathbf{S}_1 \mid g_1(x) \in \mathrm{im}(p_1 \Delta'(C_2))\}$ is compact since $p_1(\Delta'(C_2))$ is compact and g_1 is proper. But $\Delta' \pitchfork (g_1 \times g_2) \subset C_2 \times D \times \mathbf{S}_2$ is a closed subset of a compact space and so is compact. It is a zero-dimensional stratifold and oriented. We consider the sum of the orientations of this stratifold, where we recall that we equipped $\mathbf{S}_1 \times \mathbf{S}_2$ with $(-1)^{mk}$ times the product orientation. In this way we attach to a cohomology class $[\mathbf{S}_1, g_1] \in SH^k(M)$ and a homology class $[\mathbf{S}_2, g_2] \in SH_k(M)$ an integer denoted by

$$\langle [\mathbf{S}_1, g_1], [\mathbf{S}_2, g_2] \rangle \in \mathbb{Z}.$$

A transversality argument similar to the one that was used to show that that f^* is well defined implies that this number is well defined, if we assume the same transversality condition for the bordisms.

This construction gives a bilinear map which we call the **Kronecker pairing** or **Kronecker product**:

$$\langle \cdot , \cdot \rangle : SH^k(M) \times SH_k(M) \to \mathbb{Z}.$$

If M is a compact m-dimensional smooth manifold there is the following relation between the cup product, Poincaré duality and the Kronecker pairing:

Proposition 14.4. *Let $[\mathbf{S}_1, g_1] \in SH^k(M)$ and $[\mathbf{S}_2, g_2] \in SH^{m-k}(M)$ be cohomology classes. Then*

$$\langle [\mathbf{S}_1, g_1], P([\mathbf{S}_2, g_2]) \rangle = \langle [\mathbf{S}_1, g_1] \smile [\mathbf{S}_2, g_2], [M] \rangle.$$

This useful identity follows from the definitions.

The Kronecker pairing gives a homomorphism

$$SH^k(M) \to \mathrm{Hom}\,(SH_k(M), \mathbb{Z})$$

by mapping $[\mathbf{S}_1, g_1] \in SH^k(M)$ to the homomorphism assigning to $[\mathbf{S}_2, g_2] \in SH_k(M)$ the outcome of the Kronecker pairing $\langle [\mathbf{S}_1, g_1], [\mathbf{S}_2, g_2] \rangle$. We call this the **Kronecker homomorphism**:

$$\kappa : SH^k(M) \to \mathrm{Hom}(SH_k(M), \mathbb{Z}).$$

The Kronecker homomorphism from $SH^k(M)$ to $\mathrm{Hom}\,(SH_k(M), \mathbb{Z})$ commutes with induced maps $f : N \to M$:

$$\langle f^*([\mathbf{S}_1, g_1]), [\mathbf{S}_2, g_2] \rangle = \langle [\mathbf{S}_1, g_1], f_*([\mathbf{S}_2, g_2]) \rangle$$

for all $[\mathbf{S}_1, g_1] \in SH^k(M)$ and $[\mathbf{S}_2, g_2] \in SH_k(N)$.

The Kronecker homomorphism also commutes with the boundary operators in the Mayer-Vietoris sequence. The argument is the following. Let U and V be open subsets of M and let $M := U \cup V$. Choose a separating function $\rho : U \cup V \to \mathbb{R}$ as in the definition of δ. For $[\mathbf{S}_1, g_1] \in$

3. The Kronecker pairing

$SH^k(U \cap V)$ and $[\mathbf{S}_2, g_2] \in SH_{k-1}(U \cup V)$ we choose a common regular value s of ρg_1 and ρg_2. This gives a decomposition of $\mathbf{S}_1 = (\mathbf{S}_1)_+ \cup (\mathbf{S}_1)_-$ and $\mathbf{S}_2 = (\mathbf{S}_2)_+ \cup (\mathbf{S}_2)_-$ as in the definition of δ in Chapter 12, §3. Then $\delta([\mathbf{S}_1, g_1]) = [\partial(\mathbf{S}_1)_+, g_1|_{\partial(\mathbf{S}_1)_+}]$ and $d([\mathbf{S}_2, g_2]) = [\partial(\mathbf{S}_2)_+, g_2|_{\partial(\mathbf{S}_2)_+}]$. We consider the oriented regular stratifold $(\mathbf{S}_1)_+ \times (\mathbf{S}_2)_+$ with boundary $(\partial(\mathbf{S}_1)_+ \times (\mathbf{S}_2)_+) \cup_{\partial(\mathbf{S}_1)_+ \times \partial(\mathbf{S}_2)_+} -((\mathbf{S}_1)_+ \times \partial((\mathbf{S}_2)_+))$. (The product of two bounded stratifolds has, like the product of two bounded smooth manifolds, corners. There is a standard method for smoothing the corners which is based on collars. Thus the same can be done for stratifolds. Smoothing of corners is explained in a different context in appendix A.) Now we approximate the diagonal map $\Delta : X \to X \times X$ by a map Δ' which is transverse to $g_1 \times g_2 : (\mathbf{S}_1)_+ \times (\mathbf{S}_2)_+ \to X \times X$ and to the restrictions of $g_1 \times g_2$ to $\partial((\mathbf{S}_1)_+ \times (\mathbf{S}_2)_+)$ and to $\partial(\mathbf{S}_1)_+ \times \partial(\mathbf{S}_2)_+$. We consider the bounded stratifold $(\mathbf{S}_1 \times \mathbf{S}_2, g_1 \times g_2) \pitchfork (X, \Delta')$. This is a 1-dimensional stratifold with boundary $(\mathbf{S}_1 \times \mathbf{S}_2, g_1 \times g_2)|_{\partial((\mathbf{S}_1)_+ \times (\mathbf{S}_2)_+)} \pitchfork (X, \Delta')$. Since Δ' is transverse to $\partial(\mathbf{S}_1)_+ \times \partial(\mathbf{S}_2)_+$ the dimension of $(\partial(\mathbf{S}_1)_+ \times \partial(\mathbf{S}_2)_+, g_1|_{\partial(\mathbf{S}_1)_+} \times g_2|_{\partial(\mathbf{S}_2)_+}) \pitchfork (X, \Delta')$ is -1, implying that the boundary of $(\mathbf{S}_1 \times \mathbf{S}_2, g_1 \times g_2) \pitchfork (X, \Delta')$ is

$$(\partial(\mathbf{S}_1)_+ \times \mathbf{S}_2, g_1|_{\partial(\mathbf{S}_1)_+} \times g_2) \pitchfork (X, \Delta')$$
$$+(-\mathbf{S}_1 \times \partial(\mathbf{S}_2)_+, g_1 \times g_2|_{\partial(\mathbf{S}_2)_+}) \pitchfork (X, \Delta')$$

(the sign comes from the sign in the decomposition of the boundary of $(\mathbf{S}_1)_+ \times (\mathbf{S}_2)_+$). The number of oriented intersection points of $(\partial(\mathbf{S}_1)_+ \times \mathbf{S}_2, g_1|_{\partial(\mathbf{S}_1)_+} \times g_2) \pitchfork (X, \Delta')$ is the Kronecker pairing of $\delta([\mathbf{S}_1, g_1])$ and $[\mathbf{S}_2, g_2]$. The number of oriented intersection points of $(\mathbf{S}_1 \times \partial(\mathbf{S}_2)_+, g_1 \times g_2|_{\partial(\mathbf{S}_2)_+}) \pitchfork (X, \Delta')$ is the Kronecker pairing of $[\mathbf{S}_1, g_1]$ and $d([\mathbf{S}_2, g_2])$. Thus these two numbers agree.

These considerations imply that the Kronecker homomorphism gives a natural transformation

$$\kappa : SH^k(M) \to \mathrm{Hom}(SH_k(M), \mathbb{Z}).$$

Unfortunately $\mathrm{Hom}(SH_k(M), \mathbb{Z})$ is not a cohomology theory. The reason is that if $A \to B \to C$ is an exact sequence then in general the induced sequence $\mathrm{Hom}(C, \mathbb{Z}) \to \mathrm{Hom}(B, \mathbb{Z}) \to \mathrm{Hom}(A, \mathbb{Z})$ is not exact. But by a similar argument as for taking the tensor product with $\mathbb{Q}$ the induced sequence $\mathrm{Hom}(C, \mathbb{Q}) \to \mathrm{Hom}(B, \mathbb{Q}) \to \mathrm{Hom}(A, \mathbb{Q})$ is exact. Thus $\mathrm{Hom}(SH_k(X), \mathbb{Q})$ is a cohomology theory. We call the corresponding Kronecker homomorphism

$$\kappa_\mathbb{Q} : SH^k(M) \to \mathrm{Hom}(SH_k(M), \mathbb{Q}).$$

In a similar way we can define the Kronecker pairing for the $\mathbb{Z}/2$-(co)homology groups of a (not necessarily orientable) manifold M. The only difference is that we have to take the number of points mod 2 in the transverse intersection instead of the sum of the orientations as before. From the Kronecker product we obtain as before a natural transformation

$$\kappa_{\mathbb{Z}/2} : SH^k(M; \mathbb{Z}/2) \to \operatorname{Hom}(SH_k(M; \mathbb{Z}/2), \mathbb{Z}/2),$$

where now both sides are cohomology theories.

For M a point both these natural transformations are obviously isomorphisms. Thus we obtain from Proposition 14.1:

Theorem 14.5. (Kronecker Theorem) *For all smooth oriented manifolds M admitting a finite good atlas, the Kronecker homomorphism is an isomorphism:*

$$\kappa_{\mathbb{Q}} : SH^k(M; \mathbb{Q}) \cong \operatorname{Hom}(SH_k(M), \mathbb{Q})$$

and if M is not oriented:

$$\kappa_{\mathbb{Z}/2} : SH^k(M; \mathbb{Z}/2) \cong \operatorname{Hom}(SH_k(M; \mathbb{Z}/2), \mathbb{Z}/2).$$

In particular this theorem applies to all compact oriented manifolds. There is also a version of the Kronecker Theorem for integral cohomology, but the Kronecker homomorphism is not in general an isomorphism. It is still surjective and the kernel is isomorphic to the torsion subgroup of $SH_{k-1}(M)$. We will not give a proof of this result. One way to prove it is to use the isomorphism between our (co)homology groups and the classical groups defined using chain complexes. This will be explained in chapter 20. The world of chain complexes is closely related to homological algebra and in this context the integral Kronecker Theorem is rather easy to prove, as a special case of the Universal Coefficient Theorem for cohomology. One can also give a more direct proof using linking numbers, but this would lead us too far from our present context.

As announced before, we want to combine Poincaré duality and the Kronecker Theorem for closed (oriented) manifolds to obtain further relations between their homology and cohomology groups. We now present an example of this, and give an immediate application.

If we compose the Kronecker isomorphism with Poincaré duality we obtain the following non-trivial consequence:

Corollary 14.6. *Let M be a closed smooth oriented m-dimensional manifold. Then the composition of Poincaré duality with the Kronecker isomorphism induces an isomorphism:*

$$SH_{m-k}(M; \mathbb{Q}) \cong \operatorname{Hom}(SH_k(M), \mathbb{Q}).$$

Similarly if M is not necessarily oriented:
$$SH_{m-k}(M;\mathbb{Z}/2) \cong \mathrm{Hom}\,(SH_k(M;\mathbb{Z}/2),\mathbb{Z}/2).$$

An important consequence of this result is that the Euler characteristic of an odd-dimensional closed smooth manifold M vanishes. This is because the Betti numbers $b_k(M;\mathbb{Z}/2)$ are equal to $b_{m-k}(M;\mathbb{Z}/2)$ and so $(-1)^k b_k(M;\mathbb{Z}/2) + (-1)^{m-k} b_{m-k}(M;\mathbb{Z}/2) = 0$.

Corollary 14.7. *The Euler characteristic of a smooth closed odd-dimensional manifold vanishes.*

We earlier quoted a result from differential topology that there is a nowhere vanishing vector field on a closed smooth manifold if and only if the Euler characteristic vanishes. As a consequence of the corollary, we conclude that each closed odd-dimensional smooth manifold has a nowhere vanishing vector field.

4. Exercises

(1) Let $f : S^2 \to T^2$ be a continuous map where S^2 is the sphere and T^2 is the torus. Show that the map $f_* : SH_k(S^2;\mathbb{Z}/2) \to SH_k(T^2;\mathbb{Z}/2)$ is an isomorphism for $k = 0$ and the zero map for $k > 0$.

(2) Show that the spaces $\mathbb{CP}^2$ and $S^2 \vee S^4$ have the same integral homology. Are they homotopy equivalent? (Hint: Replace $S^2 \vee S^4$ by a non-compact homotopy equivalent smooth manifold.)

(3) Consider the quotient map $f : S^3 \to S^2$ when we consider S^3 as the unit sphere in $\mathbb{C}^2$ and S^2 as the Riemann sphere $\mathbb{CP}^1$. Show that $f_* : \widetilde{SH}_k(S^3;\mathbb{Z}) \to \widetilde{SH}_k(S^2;\mathbb{Z})$ is an isomorphism for $k = 0$ and the zero map for $k > 0$ but f is not null homotopic. (Hint: Show that if this map is null homotopic then $\mathbb{CP}^2$ is homotopy equivalent to $S^2 \vee S^4$.)

(4) Compute all cup products in the cohomology rings $H^*(L_k;\mathbb{Z}/2)$ and $H^*(L_k)$ of the lens spaces L_k.

(5) Let $\alpha \in SH^k(M)$ and $\beta \in H^l(N)$, show that
$$p_1^*(\alpha) \cup p_2^*(\beta) = \alpha \times \beta.$$
Hint: Show that $p_1^*(\alpha) = \alpha \times 1$ and $p_2^*(\beta) = 1 \times \beta$.

(6) Prove that the $\times$-product is natural, i.e., for a smooth map $f : M' \to M$ and $g : N' \to N$ and $\alpha \in SH^k(M)$ and $\beta \in SH^\ell(N)$ we have:
$$(f \times g)^*(\alpha \times \beta) = f^*(\alpha) \times g^*(\beta).$$

Chapter 15

The signature

As an application of the cup product, we define the signature of a closed smooth oriented $4k$-dimensional manifold and prove an important property of the signature. We recall from linear algebra the definition of the signature or index of a symmetric bilinear form over a finite dimensional $\mathbb{Q}$-vector space

$$b : V \times V \longrightarrow \mathbb{Q}.$$

The signature $\tau(b)$ is defined to be the number of positive eigenvalues minus the number of negative eigenvalues of a matrix representation of b. Equivalently, one chooses a basis $e_1, \ldots, e_r$ of V such that $b(e_i, e_j) = 0$ for $i \neq j$ and defines $\tau(b)$ as the number of e_i with $b(e_i, e_i) > 0$ minus the number of e_j with $b(e_j, e_j) < 0$. This is independent of any choices and a fundamental algebraic invariant. If we replace b by $-b$ the signature changes its sign:

$$\tau(-b) = -\tau(b).$$

Now we define the signature of a closed smooth oriented $4k$-dimensional manifold M. We have shown in chapter 7, Theorem 7.5, that the $\mathbb{Z}/2$-homology groups of a closed manifold are finitely generated, the same argument gives this for integral homology $SH_k(M)$. Thus by Poincaré duality the cohomology group $SH^{2k}(M) \cong SH_{2k}(M)$ is finitely generated. Recall that we abbreviated the fundamental class $[M, \mathrm{id}] \in SH_{4k}(M)$ by $[M]$. The **intersection form** of M is the bilinear form

$$S(M) : SH^{2k}(M) \times SH^{2k}(M) \to \mathbb{Z}$$

mapping
$$(\alpha, \beta) \mapsto \langle \alpha \smile \beta, [M] \rangle,$$
the Kronecker pairing between $\alpha \smile \beta$ and the fundamental class. Since $\alpha \smile \beta = (-1)^{(2k)^2} \beta \smile \alpha$ the intersection form is symmetric. Thus, after taking the tensor product with $\mathbb{Q}$ (which just means that we consider the matrix representing the intersection form with respect to a basis of the free part of $SH^{2k}(M)$ as a matrix with rational entries) we can consider the signature $\tau(S(M) \otimes \mathbb{Q})$ and define the **signature** of M as
$$\tau(M) := \tau(S(M) \otimes \mathbb{Q}).$$
If the dimension of M is not divisible by 4, we set $\tau(M) = 0$. If the dimension is divisible by 4, it is an important invariant of manifolds as we will see. If we replace M by $-M$ then we only replace $[M]$ by $-[M]$ and thus S changes its sign implying
$$\tau(-M) = -\tau(M).$$

Since $SH^{2k}(S^{4k}) = 0$, the signature of spheres is zero. We have computed the cohomology ring of $\mathbb{CP}^{2k}$ and we know that $SH^{2k}(\mathbb{CP}^{2k}) = \mathbb{Z}x^k$ and that $\langle x^{2k}, [\mathbb{CP}^{2k}] \rangle = 1$. Thus we have:
$$\tau(\mathbb{CP}^{2k}) = 1.$$

The significance of the signature is demonstrated by the fact that it is bordism invariant:

Theorem 15.1. (Thom) *If a compact oriented smooth manifold M is the boundary of a compact oriented smooth c-manifold W, then its signature vanishes:*
$$\tau(M) = 0.$$

The main ingredient of the proof is the following:

Lemma 15.2. *Let W be a compact smooth oriented c-manifold of dimension $2k+1$. Let $j : \partial W \to \mathring{W}$ be the map given by $j(x) := \varphi(x, \epsilon/2)$, where φ is the collar of W. Then*
$$\ker(j_* : SH_k(\partial W) \to SH_k(\mathring{W}))$$
$$\cong \operatorname{im}(j^* : SH^k(\mathring{W}) \to SH^k(\partial W) \cong SH_k(\partial W)).$$

Proof: If $[\mathbf{S}, g] \in SH_k(\partial W)$ maps to 0 under j_* there is a compact regular c-stratifold $\mathbf{T}$ with $\partial \mathbf{T} = \mathbf{S}$ and a map $G : \mathbf{T} \to \overset{\circ}{W}$ extending $j \circ g$. Now we consider $P := \mathbf{T} \cup_{\partial \mathbf{T} \times \epsilon/2} \partial \mathbf{T} \times (0, \epsilon/2]$ and extend G to a smooth proper map $\bar{G} : P \to \overset{\circ}{W}$ in such a way that for t small enough (x, t) is mapped to $\varphi(g(x), t)$. For some fixed $\delta > 0$ we consider $j_\delta : \partial W \to \overset{\circ}{W}$ by mapping x to $\varphi(x, \delta)$. For δ small enough (so that the intersection of the image of $\mathbf{T}$ with the image of j_δ is empty) we have by construction of $[P, \bar{G}]$ that $j_\delta^*([P, \bar{G}]) = \pm[\mathbf{S}, g]$. Since j_δ is homotopic to j we have shown $\ker j_* \subset \operatorname{im} j^*$.

To show the reverse inclusion, we consider $[P, h] \in SH^k(\overset{\circ}{W})$. By Sard's Theorem h is transverse to $\varphi(\partial W, \delta)$ for some $\delta > 0$. We denote $\mathbf{S} = h \pitchfork \varphi(\partial W \times \delta)$. Then $j_\delta^*([P, h]) = [\mathbf{S}, h|_\mathbf{S}]$ and — since j_δ is homotopic to j — we have $j^*([P, h]) = [\mathbf{S}, h|_\mathbf{S}]$. To show that $j_*([\mathbf{S}, h|_\mathbf{S}]) = 0$ we consider $h^{-1}(\overset{\circ}{W} - (\partial W \times (0, \delta)))$. We are finished if this is a regular c-stratifold $\mathbf{T}$ with boundary $\mathbf{S}$. Namely then $(\mathbf{T}, h|_\mathbf{T})$ is a zero bordism of $(\mathbf{S}, h|_\mathbf{S})$. Now we assume that $\mathbf{S}$ has a bicollar in P. For this we have to replace P by a bordant regular stratifold as explained in Appendix B (see Lemma B.1 in the detailed proof of the Mayer-Vietoris sequence). Then it is clear that $h^{-1}(\overset{\circ}{W} - (\partial W \times (0, \delta)))$ is an oriented regular c-stratifold $\mathbf{T}$ with boundary $\mathbf{S}$ which finishes the argument.
q.e.d.

This lemma is normally obtained from the generalization of Poincaré duality to compact oriented manifolds with boundary, the Lefschetz duality Theorem. But one only needs this partial elementary information for the proof of Theorem 15.1.

Combining this lemma with the Kronecker isomorphism (which implies that after passing to rational (co)homology we have: $j^* = (j_*)^*$, where the last $*$ denotes the dual map) we conclude that for $j_* : SH_k(\partial W) \to SH_k(\overset{\circ}{W})$:

$$\operatorname{rank}(\ker j_*) = \operatorname{rank}(\operatorname{im}((j_*)^*)).$$

From linear algebra we know that $\operatorname{rank}(\operatorname{im} j_*) = \operatorname{rank}(\operatorname{im}((j_*)^*))$ and we obtain:

$$\operatorname{rank}(\ker j_*) = \operatorname{rank}(\operatorname{im} j_*)$$

and by the dimension formula:

$$\operatorname{rank}(\ker j_*) = \frac{1}{2} \operatorname{rank} SH_k(\partial W).$$

Applying Lemma 15.2 again we finally note:

$$\operatorname{rank}(\operatorname{im} j^*) = \frac{1}{2} \operatorname{rank} SH^k(\partial W).$$

As the final preparation for the proof of Theorem 15.1 we need the following observation from linear algebra. Let $b : V \times V \to \mathbb{Q}$ be a symmetric non-degenerate bilinear form on a finite dimensional $\mathbb{Q}$-vector space. Suppose that there is a subspace $U \subset V$ with $\dim U = \frac{1}{2} \dim V$ such that, for all $x, y \in U$, we have $b(x, y) = 0$. Then $\tau(b) = 0$. The reason is the following. Let $e_1, \ldots, e_n$ be a basis of U. Since the form is non-degenerate, there are elements $f_1, \ldots, f_n$ in V such that $b(f_i, e_j) = \delta_{ij}$ and one can further achieve that $b(f_i, f_j) = 0$. This implies that $e_1, \ldots, e_n, f_1, \ldots, f_n$ are linear independent and thus form a basis of V. Now consider $e_1 + f_1, \ldots, e_n + f_n, e_1 - f_1, \ldots, e_n - f_n$ and note that, with respect to this basis, b has the form

$$\begin{pmatrix} 2 & & & & & \\ & \ddots & & & & \\ & & 2 & & & \\ & & & -2 & & \\ & & & & \ddots & \\ & & & & & -2 \end{pmatrix}$$

and thus

$$\tau(b) = 0.$$

Proof of Theorem 15.1: We first note that for $\alpha \in \operatorname{im} j^*$ and $\beta \in \operatorname{im} j^*$ the intersection form $S(\partial W)(\alpha, \beta)$ vanishes. For if $\alpha = j^*(\bar{\alpha})$ and $\beta = j^*(\bar{\beta})$, then

$$S(\partial W)(\alpha, \beta) = \langle j^*(\bar{\alpha}) \smile j^*(\bar{\beta}), [\partial W] \rangle = \langle \bar{\alpha} \smile \bar{\beta}, j_*([\partial W]) \rangle = 0$$

since $j_*([\partial W]) = 0$ (note that W is a zero bordism).

Thus the intersection form vanishes on $\operatorname{im} j^*$. By Poincaré duality the intersection form $S(\partial W) \otimes \mathbb{Q}$ is non-degenerate. Since the rank of $\operatorname{im} j^*$ is $\frac{1}{2} \operatorname{rank} SH^k(\partial W)$ the proof is finished using the considerations above from linear algebra.
q.e.d.

15. The signature

The significance of Theorem 15.1 becomes more visible if we define bordism groups of compact oriented smooth manifolds. They were introduced by Thom [**Th 1**] who computed their tensor product with $\mathbb{Q}$ and provided with this the ground for very interesting applications (for example the signature theorem, which in a special case we will discuss later). The group Ω_n is defined as the bordism classes of compact oriented smooth manifolds. More precisely the elements in Ω_n are represented by a compact smooth n-dimensional manifold M and two such manifolds M and M' are equivalent if there is a compact oriented manifold W with boundary $M \sqcup (-M')$. The sum is given by disjoint union and the inverse of a bordism class $[M]$ is $[-M]$. Thus the definition is analogous to the definition of $SH_n(\text{pt})$, the difference being that we only consider manifolds instead of regular stratifolds.

Whereas it was simple to determine $SH_n(\text{pt})$, it is very difficult to compute the groups Ω_n. This difficulty is indicated by the following consequence of Theorem 15.1.

The signature of a disjoint union of manifolds is the sum of the signatures, and $\tau(-M) = -\tau(M)$. Thus we conclude from Theorem 15.1, that

$$\tau : \Omega_{4k}(\text{pt}) \to \mathbb{Z}$$

is a homomorphism. This homomorphism $\tau : \Omega_{4k}(\text{pt}) \to \mathbb{Z}$ is a surjective map. The reason is that $\tau(\mathbb{CP}^{2k}) = 1$.

Thus we obtain:

Corollary 15.3. *For each $k \geq 0$ the groups $\Omega_{4k}(\text{pt})$ are non-trivial.*

It is natural to ask what the signature of a product of two manifolds is. It is the product of the signatures of the two manifolds:

Theorem 15.4. *Let M and N be closed oriented smooth manifolds. Then*

$$\tau(M \times N) = \tau(M)\tau(N).$$

The proof is based on the Künneth theorem for rational cohomology and Poincaré duality and we refer to Hirzebruch's original proof [**Hir**], p. 85 or better yet, suggest that readers do the following exercise.

1. Exercises

(1) Prove Theorem 15.4. (Hint: Apply the Künneth theorem to compute the middle rational cohomology of the product. Decompose the intersection form as the orthogonal sum of the tensor product of the intersection forms of the factors and the rest. Show that the rest is the orthogonal sum of summands which contain a half rank subspace on which the form vanishes implying that all these terms have signature zero.)

(2) Show that the signature mod 2 of a closed oriented smooth manifold M is equal to the Euler characteristic mod 2 (also if the dimension is not divisible by 4).

(3) Show that the signature of the connected sum $M \# N$ is the sum of the signatures of M and N.

(4) Prove that the signature of a mapping torus M_f is zero, where M is a closed oriented smooth manifold and f an orientation-preserving diffeomorphism (see exercise 12 in chapter 8).

Chapter 16

The Euler class

1. The Euler class

We recall the definition of the Euler class. Let $p : E \to M$ be a smooth oriented k-dimensional vector bundle over a smooth oriented manifold M. Let $s : M \to E$ be the zero section. Then $e(E) := s^*[M, s] \in SH^k(M)$ is the **Euler class** of E. The Euler class is called a **characteristic class**. We will define other characteristic classes like the Chern, Pontrjagin and Stiefel-Whitney classes.

By construction the Euler classes of bundles $p : E \to M$ and $p' : E' \to M$, which are orientation-preserving isomorphic, are equal. Thus the Euler class is an invariant of the oriented isomorphism type of a smooth vector bundle. We also recall Proposition 13.3, that if a smooth oriented bundle E has a nowhere vanishing section then $e(E) = 0$. In particular the Euler class of a positive dimensional trivial bundle is 0. Finally, if we change the orientation of E and $f : E \to -E$ is the identity map, then $f^*[M, s] = [-M, s]$, which implies that, since s and f commute, $e(-E) = -e(E)$.

The following properties of the Euler class are fundamental.

Theorem 16.1. *Let $p : E \to M$ be a smooth oriented vector bundle. Then, if $-E$ is E with opposite orientation:*

$$e(-E) = -e(E).$$

If $f : N \to M$ is a smooth map, then the Euler class is natural:
$$e(f^*E) = f^*(e(E)).$$
If $q : F \to M'$ is another smooth oriented vector bundle then
$$e(E \times F) = e(E) \times e(F),$$
and if $M = M'$,
$$e(E \oplus F) = e(E) \smile e(F).$$

Here we recall that the Whitney sum $E \oplus F := \Delta^*(E \times F)$ is the pullback of $E \times F$ under the diagonal map. The fibre of $E \oplus F$ at x is $E_x \oplus F_x$.

Proof: The first property follows from the definition of the Euler class. For the second property we divide it up into a series of cases which are more or less obvious (we suggest that the readers add details as an exercise). We first consider the case where $N \subset M$ is a submanifold of M and f is the inclusion. In this case it is clear from the definition that $e(f^*E) = f^*e(E)$. Next we assume that f is a diffeomorphism and note that the property follows again from the definition. Combining these two cases we conclude that the statement holds for embeddings $f : N \to M$. A next obvious case is given by considering for an arbitrary manifold N the projection $p : M \times N \to M$ and seeing that $e(p^*(E)) = p^*(e(E))$. Now we consider the general case of a smooth map $f : N \to M$. Let $g : N \to M \times N$ be the map $x \mapsto (f(x), x)$. This is an embedding and $pg = f$. Thus from the cases above we see:

$$e(f^*(E)) = e((pg)^*(E)) = e(g^*(p^*(E))) = g^*(e(p^*(E)))$$
$$= g^*(p^*(e(E))) = (pg)^*e((E)) = f^*(e(E)).$$

The property $e(E \times F) = e(E) \times e(F)$ follows again from the definition. Combining this with the definition of the Whitney sum and naturality we conclude $e(E \oplus F) = e(E) \smile e(F)$.
q.e.d.

The following is a useful observation.

Corollary 16.2. *Let $p : E \to M$ be a smooth oriented vector bundle. If E is odd-dimensional, then*
$$2e(E) = 0.$$

Proof: If E is odd-dimensional $-\mathrm{id} : E \to E$ is an orientation-reversing bundle isomorphism and thus we conclude that $e(E) = -e(E)$.
q.e.d.

2. Euler classes of some bundles

Remark: The name "Euler class" was chosen since there is a close relation between the Euler class of a closed oriented smooth manifold M and the Euler characteristic. Namely:
$$e(M) = \langle e(TM), [M] \rangle,$$
the Euler characteristic is the Kronecker product between the Euler class of the tangent bundle and the fundamental class of M. By definition of the Euler class and the Kronecker product this means that if $v : M \to TM$ is a section, which is transverse to the zero section, then the Euler characteristic is the sum of the orientations of the intersections of v with the zero section. This identity is the Poincaré-Hopf Theorem.

In special cases one can compute $\langle e(TM), [M] \rangle$ directly and verify the Poincaré-Hopf Theorem. We have done this already for spheres. For complex projective spaces one has:
$$\langle e(T\mathbb{CP}^m), [\mathbb{CP}^m] \rangle = m+1$$
We leave this as an exercise to the reader. Combining it with Proposition 9.5 we conclude:

Theorem 16.3. *Each vector field on $\mathbb{CP}^n$ has a zero.*

2. Euler classes of some bundles

Now we compute the Euler class of some bundles. As a first example we consider the **tautological bundle**
$$p : L = \{([x], v) \in \mathbb{CP}^n \times \mathbb{C}^{n+1} \mid v \in \mathbb{C} \cdot x\} \to \mathbb{CP}^n.$$
This is a complex vector bundle of complex dimension 1, whose fibre over $[x]$ is the vector space generated by x. By construction the restriction of the tautological bundle over $\mathbb{CP}^n$ to $\mathbb{CP}^k$ for some $k < n$ is the tautological bundle over $\mathbb{CP}^k$. This is the reason which allows us, by abuse of notation, to use the same name for bundles over different spaces. A complex vector space V considered as a real vector space has a canonical orientation. Namely choose a basis $(v_1, \ldots, v_n)$ and consider the basis of the real vector space $(v_1, iv_1, v_2, iv_s, \ldots, v_n, iv_n)$. The orientation given by this basis is independent of the choice of the basis $(v_1, \ldots, v_n)$ (why?). Using this orientation fibrewise we can consider L as a 2-dimensional oriented real vector bundle. To compute the Euler class we first consider the case $p : L \to \mathbb{CP}^1$ and consider the section given by
$$s : [x_0, x_1] \mapsto ([x_0, x_1], x_0 \cdot \bar{x}_1, x_1 \cdot \bar{x}_1).$$

Then
$$s([x_0, x_1]) = 0 \Leftrightarrow x_1 = 0.$$
To check whether the section is transverse to the zero section and to compute $\epsilon([1,0])$ (the sign coming from the orientations at this point), we choose local coordinates around this point: $\varphi : \{([x], v) \in \mathbb{CP}^1 \times \mathbb{C}^2 \mid v \in \mathbb{C}x$ and $x_0 \neq 0\} \to \mathbb{C} \times \mathbb{C}$ mapping $([x], v)$ to $(\frac{x_1}{x_0}, \mu)$, where $v = \mu(1, \frac{x_1}{x_0})$. This map is an isomorphism. With respect to this trivialization, we have $p_2 \varphi s([1, x_1]) = p_2 \varphi([1, x_1], (\bar{x}_1, x_1 \cdot \bar{x}_1)) = \bar{x}_1$. Thus s is transverse to the zero section and $\epsilon([1, 0]) = -1$. We conclude:

Proposition 16.4. $\langle e(L), [\mathbb{CP}^1] \rangle = -1$.

We return to the tautological bundle $p : L \to \mathbb{CP}^n$ over $\mathbb{CP}^n$. The restriction of $p : L \to \mathbb{CP}^n$ to $\mathbb{CP}^1$ is $p : L \to \mathbb{CP}^1$. Using the naturality of the Euler class the statement above implies $\langle e(L), [\mathbb{CP}^1, i] \rangle = -1$. We recall that we defined $x := [\mathbb{CP}^{n-1}, i] \in SH^2(\mathbb{CP}^n)$ and showed that $\langle x, [\mathbb{CP}^1, i] \rangle = 1$. Thus Proposition 16.4 implies:
$$e(L) = -x.$$

As another example we consider the complex line bundle
$$E_k := D^2 \times \mathbb{C} \cup_{f_k} -D^2 \times \mathbb{C} \xrightarrow{p_1} D^2 \cup -D^2 = S^2,$$
where $f_k : S^1 \times \mathbb{C} \to S^1 \times \mathbb{C}$ maps $(z, v) \mapsto (z, z^k \cdot v)$. This bundle is closely related to lens spaces. If we equip E_k with the Riemannian metric induced from the standard Euclidean metric on $\mathbb{C} = \mathbb{R}^2$, the lens space L_k is the sphere bundle SE_k. The bundle E_k can naturally be equipped with the structure of a smooth vector bundle by describing it as:
$$\mathbb{C} \times \mathbb{C} \cup_{g_k} \mathbb{C} \times \mathbb{C}$$
with
$$g_k : \mathbb{C}^* \times \mathbb{C} \longrightarrow \mathbb{C}^* \times \mathbb{C}$$
$$(x, y) \longmapsto (1/x, x^k y).$$
If we consider E_k above as an oriented bundle over $D^2 \cup_{\bar{z}} D^2$ instead of over the diffeomorphic oriented manifold $D^2 \cup -D^2$, we have to describe $E_k = D^2 \times \mathbb{C} \cup_{f'_k} D^2 \times \mathbb{C} \xrightarrow{p_1} D^2 \cup_{\bar{z}} D^2 = S^2$, where $f'_k(z, v) = (\bar{z}, z^k \cdot v) = (1/z, z^k \cdot v)$.

This describes E_k as a smooth (even holomorphic) vector bundle over $\mathbb{C} \cup_{\frac{1}{x}} \mathbb{C} = S^2$. Now we first compute $\langle e(E_1), [S^2] \rangle$ by choosing a section which is transverse to the zero section. For $||x|| < 2$ and $x \in \mathbb{C}$, the first copy of $\mathbb{C}$ in $\mathbb{C} \cup_{\frac{1}{x}} \mathbb{C}$, we define the section as $s(x) := (x, \bar{x})$ and for z in the second copy we define $s(z) := (z, \rho(||z||)^2)$, where $\rho : [0, \infty) \to (0, \infty)$

2. Euler classes of some bundles

is a smooth function with $\rho(s) = 1/s$ for $s > 1/2$. This smooth section has a single zero at 0 in the first summand and there it intersects transversely with local orientation -1.

We conclude:
$$\langle e(E_1), [S^2] \rangle = -1.$$

From this we compute $\langle e(E_k), [S^2] \rangle$ for all k by showing
$$\langle e(E_{k+\ell}), [S^2] \rangle = \langle e(E_k), [S^2] \rangle + \langle e(E_\ell), [S^2] \rangle.$$

Consider D^3 with two holes as in the following picture, and denote this 3-dimensional oriented manifold by M:

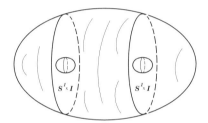

Decompose M along the two embedded $S^1 \times I$'s and denote the three resulting areas by M_1, M_2 and M_3. Now construct a bundle over M by gluing $M_1 \times \mathbb{C}$ to $M_2 \times \mathbb{C}$ via $f_k \times \mathrm{id} : S^1 \times \mathbb{C} \times I \to S^1 \times \mathbb{C} \times I$, and $M_2 \times \mathbb{C}$ to $M_3 \times \mathbb{C}$ via $f_\ell \times \mathrm{id} : S^1 \times \mathbb{C} \times I \to S^1 \times \mathbb{C} \times I$ to obtain
$$E := M_1 \times \mathbb{C} \cup_{f_k \times \mathrm{id}} M_2 \times \mathbb{C} \cup_{f_\ell \times \mathrm{id}} M_3 \times \mathbb{C} \xrightarrow{p_1} M_1 \cup M_2 \cup M_3 = M.$$

Orient M so that $\partial M = S^2 + (-S_1^2) + (-S_2^2)$, where S_i^2 are the boundaries of the two holes. Then the reader should convince himself that
$$E|_{S^2} = E_{k+\ell}$$
since we can combine the two gluings by f_k and f_ℓ along the two circles into one gluing by $f_\ell \circ f_k = f_{\ell+k}$. By construction, $E|_{S_1^2} = E_k$ and $E|_{S_2^2} = E_\ell$. Next we note that
$$\langle e(E), [\partial M] \rangle = 0$$
since $[\partial M]$ is zero in $SH_2(M)$ (M itself is a zero bordism of ∂M). But
$$\langle e(E), [\partial M] \rangle = \langle e(E), ([S^2] + [-S_1^2] + [-S_2^2]) \rangle$$
$$= \langle e(E|_{S^2}), [S^2] \rangle - \langle e(E|_{S_1^2}), [S_1^2] \rangle - \langle e(E|_{S_2^2}), [S_2^2] \rangle$$
$$= \langle e(E_{k+\ell}), [S^2] \rangle - \langle e(E_k), [S^2] \rangle - \langle e(E_l), [S^2] \rangle.$$

Since $\langle e(E), [\partial M]\rangle = 0$ we have shown:

Lemma 16.5. *The map $\mathbb{Z} \to \mathbb{Z}$ mapping k to $\langle e(E_k), [S^2]\rangle$ is a homomorphism.*

Combining this with the fact $\langle e(E_1), [S^2]\rangle = -1$, we conclude

Proposition 16.6.
$$\langle e(E_k), [S^2]\rangle = -k.$$
In particular: There is an orientation-preserving bundle isomorphism between E_k and E_r if and only if $k = r$.

In complete analogy we study the bundle $E_{k,\ell}$ over S^4 given as
$$D^4 \times \mathbb{H} \cup_{f_{k,\ell}} -D^4 \times \mathbb{H} \xrightarrow{p_1} D^4 \cup -D^4 = S^4$$
where
$$f_{k,\ell}(z, v) = (z, z^k \cdot v \cdot z^\ell)$$
and we use quaternionic multiplication ($z \in S^3$). As in the case of E_k over S^4, we show that
$$\langle e(E_{1,0}), [S^4]\rangle = -1.$$

By the same argument as in the case of E_k, one shows
$$\langle e(E_{k+k', l+\ell'}), [S^4]\rangle = \langle e(E_{k,\ell}), [S^4]\rangle + \langle e(E_{k',\ell'}), [S^4]\rangle$$
or, in other words, that the map $\mathbb{Z} \times \mathbb{Z} \to \mathbb{Z}$ mapping (k, ℓ) to $\langle e(E_{k,\ell}), [S^4]\rangle$ is a homomorphism.

Next we consider the following isomorphism of $\mathbb{H}$, considered as a real vector space:
$$(z_1, z_2) \mapsto (\bar{z}_1, -z_2) =: \overline{(z_1, z_2)}$$
and note that, for $z \in S^3$, we have $\bar{z} = z^{-1}$. Further $\overline{x \cdot y} = \bar{y} \cdot \bar{x}$. Now consider the bundle isomorphism
$$E_{k,\ell} \to E_{-\ell, -k}$$
mapping $(x, v) \mapsto (x, \bar{v})$. Since $v \mapsto \bar{v}$ is orientation-reversing, this implies
$$E_{k,\ell} \cong -E_{-\ell,-k}$$
and so
$$-\langle e(E_{k,\ell}), [S^4]\rangle = \langle e(E_{-\ell,-k}), [S^4]\rangle.$$

This implies
$$\langle e(E_{k,\ell}), [S^4]\rangle = c(k+\ell)$$
for some constant c. Since $\langle e(E_{1,0}), [S^4]\rangle = -1$, we conclude $c = -1$ and thus we have shown

Proposition 16.7. $\langle e(E_{k,\ell}), [S^4]\rangle = -k - \ell$.

3. The top Stiefel-Whitney class

If we consider n-dimensional smooth vector bundles E which are not necessarily oriented over not necessarily oriented manifolds M we can define the class $w_n(E) \in SH^n(M; \mathbb{Z}/2)$ as $s^*([M,s])$. It is called the **n-th Stiefel-Whitney class of E** or the **top Stiefel-Whitney class**. Perhaps a better name for the top Stiefel-Whitney class would be to call it the mod 2 Euler class, since it is the version of the Euler class for $\mathbb{Z}/2$-cohomology. The "n" indicates that there are other Stiefel-Whitney classes $w_k(E) \in SH^k(M; \mathbb{Z}/2)$, which is the case. They are treated in the next chapter. These classes are zero for $k > n$, which is why we call $w_n(E)$ the top Stiefel-Whitney class. It has properties analogous to the Euler class. If E and M are oriented then the top Stiefel-Whitney class is the Euler class considered (by reduction mod 2) as a class in $\mathbb{Z}/2$-cohomology.

4. Exercises

(1) Let E and F be n-dimensional oriented smooth vector bundles over n-dimensional closed smooth oriented manifolds M and N. Construct a smooth oriented vector bundle $E\#F$ over $M\#N$ such that the bundle agrees outside the discs used to construct the connected sum with E and F and $\langle e(E\#F), [M\#N]\rangle = \langle e(E), [M]\rangle + \langle e(F), [N]\rangle$.

(2) Construct for each integer k an oriented smooth 2-dimensional vector bundle E over a surface F_g of genus g such that $\langle e(E), [F_g]\rangle = k$.

(3) Let E be a complex line bundle (the complex dimension of E is 1) over M. Let v be a non-zero vector in E_x. Show that the basis (v, iv) determines a well defined orientation of E_x (independent of the choice of v) and that this makes E an oriented real vector bundle of real dimension 2. Let F be another complex line bundle. Show that $e(E \otimes F) = e(E) + e(F)$, where $E \otimes F$ is the vector bundle obtained by taking fibrewise the tensor product. (Hint: Consider the vector bundle $p_1^*(E) \otimes p_2^*(F)$ over $M \times M$.)

(4) Let E be a 2-dimensional vector bundle over S^n. Show that E is trivial if $n > 2$. (Hint: You can use that the $\pi_i(SO(2)) = 0$ for $i > 1$.)

(5) Let E be a vector bundle over a simply connected CW-complex X. Show that E is orientable.

(6) Let M be an n-dimensional smooth manifold with an n-dimensional oriented smooth vector bundle E over it such that $E \oplus M \times \mathbb{R}$ is isomorphic to $M \times \mathbb{R}^{n+1}$. Show that E is trivial if and only if $e(E) = 0$. You are allowed to use that $E|_{M-\text{pt}}$ is trivial and that the statement holds for $M = S^n$.

Chapter 17

Chern classes and Stiefel-Whitney classes

Now we define the Chern classes of a complex vector bundle $p : E \to M$ over a smooth oriented manifold M. We remind the reader that a smooth **k-dimensional complex vector bundle** is a smooth map $p : E \to M$ together with a $\mathbb{C}$-vector space structure on the fibres which is locally isomorphic to $U \times \mathbb{C}^k$, where "isomorphism" means diffeomorphism and fibrewise $\mathbb{C}$-linear. For example we know that the tautological bundle $p : L \to \mathbb{CP}^n$ is a 1-dimensional complex vector bundle. If E and F are complex vector bundles the Whitney sum $E \oplus F$ is a complex vector bundle. Given two complex vector bundles E and F one can consider their tensor product $E \otimes_{\mathbb{C}} F$ which is obtained by taking fibrewise the tensor product to obtain a new complex vector bundle [**Mi-St**]. If E and F are smooth vector bundles then $E \otimes_{\mathbb{C}} F$ is again smooth.

To prepare for the definition of the Chern classes we consider, for a smooth manifold M, the homology of $M \times \mathbb{CP}^N$, for some N. By the Künneth Theorem and the fact that $SH^*(\mathbb{CP}^N) = \mathbb{Z}[e(L)]/_{e(L)^{N+1}}$ (implying that the cohomology of $\mathbb{CP}^N$ is torsion-free) we have for $k \leq N$ (if M admits a finite good atlas):

$$SH^k(M \times \mathbb{CP}^N) \cong (SH^k(M) \otimes \mathbb{Z} \cdot 1) \oplus (SH^{k-2}(M) \otimes \mathbb{Z} \cdot e(L))$$
$$\oplus (SH^{k-4}(M) \otimes \mathbb{Z} \cdot (e(L) \smile e(L))) \oplus \cdots.$$

Actually the same result is true for arbitrary manifolds M as one can show inductively over N using the Mayer-Vietoris sequence. Now let $p : E \to M$ be a smooth k-dimensional complex vector bundle and consider $p_1^* E \otimes_{\mathbb{C}} p_2^* L$, a complex vector bundle over $M \times \mathbb{CP}^N$ for some $N \geq k$, where p_1 and p_2 are the projections to the first and second factor. Since every k-dimensional complex vector bundle considered as a real bundle has a canonical orientation, we can consider the Euler class $e(p_1^* E \otimes_{\mathbb{C}} p_2^* L) \in SH^{2k}(M \times \mathbb{CP}^N)$. Using the isomorphism above we define the **Chern classes** $c_i(E) \in SH^{2i}(M)$ by the equation

$$e(p_1^* E \otimes_{\mathbb{C}} p_2^* L) = \sum_{i=0}^{k} c_i(E) \times e(L)^{k-i}.$$

In other words the Chern classes are the coefficients of $e(p_1^* E \otimes_{\mathbb{C}} p_2^* L)$ if we consider the Euler class as a "polynomial" in $e(L)$.

Since for the inclusion $i : \mathbb{CP}^N \to \mathbb{CP}^{N+1}$ we know that $i^* L$ is the tautological bundle over $\mathbb{CP}^N$, if L was the tautological bundle over $\mathbb{CP}^{N+1}$, this definition does not depend on N for $N \geq k$.

We prove some basic properties of the Chern classes. The naturality of the Euler class implies that the Chern classes are natural, i.e., if $f : N \to M$ is a smooth map, then

$$c_k(f^*(E)) = f^*(c_k(E)).$$

The Chern classes depend only on the isomorphism class of the bundle. Both these facts imply that the Chern classes of a trivial bundle are zero except $c_0 = 1$. By restricting the bundle to a point we conclude that for arbitrary bundles E we have

$$c_0(E) = 1.$$

By construction $c_i(E) = 0$ for $i > k$, where k is the complex dimension of E. Next we note that $c_k(E) = e(E)$. To see this, fix a point $x_0 \in \mathbb{CP}^N$ and consider the inclusion

$$\begin{aligned} j: M &\longrightarrow M \times \mathbb{CP}^N \\ x &\longmapsto (x, x_0). \end{aligned}$$

Then $j^*(p_1^* E \otimes_{\mathbb{C}} p_2^* L) = j^*(p_1^* E) \otimes_{\mathbb{C}} j^*(p_2^* L) \cong j^*(p_1^* E) = E$, since $p_2 j$ is the constant map and so $j^*(p_2^* L)$ is the product bundle $M \times \mathbb{C}$. On the other hand $j^* : SH^{2k}(M \times \mathbb{CP}^N) \to SH^{2k}(M)$ maps $SH^{2k}(M) \otimes \mathbb{Z} \cdot e(L)^0 \cong SH^{2k}(M)$ identically to $SH^{2k}(M)$ and the other summands in the decomposition to 0. Thus $e(E) = e(j^*(p_1^* E \otimes_{\mathbb{C}} p_2^* L)) = j^*(e(p_1^* E \otimes_{\mathbb{C}} p_2^* L)) = j^*(c_k(E)) \times e(L)^0$, and therefore

$$e(E) = c_k(E).$$

This property together with the following product formula is a basic feature of the Chern classes. We would like to know $c_r(E \oplus F)$ for k and ℓ-dimensional complex vector bundles E and F over M. For this we choose $N \geq k + \ell$ and note that
$$p_1^*(E \oplus F) \otimes_{\mathbb{C}} p_2^* L = (p_1^* E \otimes_{\mathbb{C}} p_2^* L) \oplus (p_1^* F \otimes_{\mathbb{C}} p_2^* L).$$
Then we conclude from
$$e((p_1^* E \otimes_{\mathbb{C}} p_2^* L) \oplus (p_1^* F \otimes_{\mathbb{C}} p_2^* L)) = e(p_1^* E \otimes_{\mathbb{C}} p_2^* L) \smile e(p_1^* F \otimes_{\mathbb{C}} p_2^* L)$$
and the definition of the Chern classes:
$$\sum_{i=0}^{k+\ell} c_i(E \oplus F) \times e(L)^{k+\ell-i} = (\sum_{r=0}^{k} c_r(E) \times e(L)^{k-r}) \smile (\sum_{s=0}^{\ell} c_s(F) \times e(L)^{\ell-s}),$$
that
$$c_i(E \oplus F) = \sum_{r+s=i} c_r(E) \smile c_s(F).$$
A convenient way to write the product formula is to consider the Chern classes as elements of the cohomology ring $SH^*(M) = \bigoplus_k SH^k(M)$. We define the **total Chern class** as
$$c(E) := \sum_k c_k(E) \in SH^*(M).$$
Then the product formula translates to:
$$c(E \oplus F) = c(E) \smile c(F).$$
We summarize these properties as

Theorem 17.1. *Let E be a k-dimensional smooth complex vector bundle over M.*
- *The Chern classes are natural, i.e., if $f : N \to M$ is a smooth map, then*
$$c_k(f^*(E)) = f^*(c_k(E)).$$
- *The Chern classes depend only on the isomorphism type of the bundle.*
- *For $i > k$ we have*
$$c_i(E) = 0.$$
- *For $i = 0$ resp. k we have*
$$c_0(E) = 1,$$
$$c_k(E) = e(E),$$
in particular $c_1(L) = -x$, where L and x are as in Proposition 16.4.
- *If E and F are smooth complex vector bundles over M, then* (**Whitney formula**)
$$c_r(E \oplus F) = \sum_{i+j=r} c_i(E) \smile c_j(F),$$

or using the total Chern class:

$$c(E \oplus F) = c(E) \smile c(F).$$

One can show that these properties characterize the Chern classes uniquely [**Mi-St**].

We conclude this chapter by briefly introducing Stiefel-Whitney classes (although we will not apply them in this book). The definition is completely analogous to the definition of the Euler class and the Chern classes. The main difference is that we will replace oriented or even complex vector bundles by arbitrary vector bundles.

If E is a k-dimensional vector bundle (not oriented) we made the same construction as for the Euler class with $\mathbb{Z}/2$-cohomology instead of integral cohomology and defined the highest Stiefel-Whitney class $w_k(E) := s^*[M, s] \in SH^k(M; \mathbb{Z}/2)$. This class fulfills the analogous properties that were shown for the Euler class in Theorem 16.1.

Now we define the lower Stiefel-Whitney classes. This is done in complete analogy to the Chern classes, where we replace the Euler class by the k-th Stiefel-Whitney class and the tautological bundle over the complex projective space by the tautological bundle L over $\mathbb{R}\mathbb{P}^N$. This is a 1-dimensional real bundle. The $\mathbb{Z}/2$-cohomology of $M \times \mathbb{R}\mathbb{P}^N$ is:

$$SH^k(M \times \mathbb{R}\mathbb{P}^N; \mathbb{Z}/2) \cong (SH^k(M; \mathbb{Z}/2) \otimes \mathbb{Z}/2) \cdot 1$$
$$\oplus (SH^{k-1}(M; \mathbb{Z}/2) \otimes \mathbb{Z}/2) \cdot w_1(L)$$
$$\oplus (SH^{k-2}(M; \mathbb{Z}/2) \otimes \mathbb{Z}/2) \cdot (w_1(L) \smile w_1(L)) \oplus \cdots.$$

Then we define the **Stiefel-Whitney classes** of a real smooth vector bundle E of dimension k over M, denoted $w_i(E) \in SH^i(M)$, by the equation

$$w_k(p_1^*(E) \otimes_{\mathbb{R}} p_2^*(L)) = \sum_{i=0}^{k} w_i(E) \times w_1(L)^{k-i}.$$

In other words, the Stiefel-Whitney classes are the coefficients of $e(p_1^*(E) \otimes_{\mathbb{R}} p_2^*(L))$ if we consider the Euler class as a "polynomial" in $w_1(L)$. By an argument similar to the one given for Theorem 17.1 one proves:

Theorem 17.2. *Let E be a k-dimensional smooth real vector bundle over M. Then analogues of the statements in the previous theorem holds. In particular for $i > k$:*

$$w_i(E) = 0.$$

If E and F are smooth vector bundles over M, then (**Whitney formula**)
$$w_r(E \oplus F) = \sum_{i+j=r} w_i(E) \smile w_j(F).$$

1. Exercises

(1) Let E be a complex vector bundle over M. Let $\bar{E}$ be the bundle with the conjugate complex structure, i.e., multiplication by λ is given by multiplication with $\bar{\lambda}$. Show that
$$c_k(\bar{E}) = (-1)^k c_k(E).$$

(2) Show that the first Chern class of the tensor product of two complex vector bundles is the sum of the first Chern classes.

(3) Compute the Chern classes of the bundle over $S^2 \times S^2$ given by $E := p_1^*(L) \oplus p_2^*(L)$, where L is the tautological bundle over $S^2 = \mathbb{CP}^1$.

(4) Construct a complex line bundle F over $S^2 \times S^2$ with first Chern class $-c_1(E)$, where E is as in the previous exercise.

(5) Show that a complex line bundle E over S^2 is trivial if and only if $c_1(L) = 0$. (You can use that the isomorphism classes of complex line bundles over S^2 are isomorphic as a set to $\pi_1(S^1)$ under the map which to each element $[f] \in \pi_1(S^1)$ attaches the bundle $D^2 \times \mathbb{C} \cup_f D^2 \times \mathbb{C}$ obtained by identifying $(x,y) \in S^1 \times \mathbb{C}$ with $(x, f(x)y)$ in the other copy.)

(6) Let E be a complex vector bundle over $S^n \times S^m$, whose restriction to $S^n \vee S^m$ is trivial. Let $p: S^n \times S^m \to S^{n+m}$ be the pinch map which collapses everything outside a small disc in $S^n \times S^m$ to a point and is the identity on the interior. Construct a bundle F over S^{n+m} such that $p^*(F)$ is isomorphic to E.

(7) Construct a complex vector bundle E over S^4 with $\langle c_2(E), [S^4] \rangle = 1$.

Chapter 18

Pontrjagin classes and applications to bordism

1. Pontrjagin classes

To obtain invariants for k-dimensional real vector bundles E we simply complexify the bundle, considering

$$E \otimes_{\mathbb{R}} \mathbb{C}.$$

This means that we replace the fibres E_x of E by the complex vector spaces $E_x \otimes_{\mathbb{R}} \mathbb{C}$ or equivalently by $E_x \oplus E_x$ with complex vector space structure given by $i \cdot (v, w) := (-w, v)$. This is a complex vector bundle of complex dimension k and we define the r-th **Pontrjagin class**

$$p_r(E) := (-1)^r c_{2r}(E \otimes_{\mathbb{R}} \mathbb{C}) \in SH^{4r}(M).$$

Here one might wonder why we have not taken $c_{2r+1}(E \otimes_{\mathbb{R}} \mathbb{C})$ into account. The reason is that these classes have order 2, as we will discuss. Also the sign convention asks for an explanation. One could leave out the sign without any problem. Probably the historical reason for the sign convention is that for $2n$-dimensional oriented bundles one can show (see exercise 2):

$$p_n(E) = e(E) \smile e(E).$$

We prepare for the argument that the classes $c_{2r+1}(E \otimes_\mathbb{R} \mathbb{C})$ are 2-torsion with some general considerations. If V is a complex k-dimensional vector space, we consider its conjugate complex vector space $\bar{V}$ with new scalar multiplication $\lambda \diamond v := \bar{\lambda} \cdot v$. Note that the orientation of $\bar{V}$, as a real vector space, is $(-1)^k$ times the orientation of V (why?). Taking the conjugate complex structure fibrewise we obtain for a complex bundle E the **conjugate bundle** $\bar{E}$. The change of orientation of vector spaces translates to complex vector bundles giving for a k-dimensional complex vector bundle that as oriented bundles $\bar{E} \cong (-1)^k E$. From this one concludes (exercise 1, chapter 17):

$$c_i(\bar{E}) = (-1)^i c_i(E).$$

Now we note that since $\mathbb{C}$ is as a complex vector space isomorphic to its conjugate this isomorphism induces an isomorphism:

$$E \otimes_\mathbb{R} \mathbb{C} \cong \overline{E \otimes_\mathbb{R} \mathbb{C}}.$$

Thus $c_{2r+1}(E \otimes_\mathbb{R} \mathbb{C}) = -c_{2r+1}(E \otimes_\mathbb{R} \mathbb{C})$ implying $2c_{2r+1}(E \otimes_\mathbb{R} \mathbb{C}) = 0$.

Since $2c_{2r+1}(E \otimes_\mathbb{R} \mathbb{C}) = 0$, the product formula for the Chern classes gives the corresponding **product formula for the Pontrjagin classes** of real vector bundles E and F:

$$p_r(E \oplus F) = \sum_{i+j=r} p_i(E) \smile p_j(F) + \beta,$$

where $2\beta = 0$.

We introduce the **total Pontrjagin class**:

$$p(E) := \sum_k p_k(E) \in SH^*(M)$$

and rewrite the product formula as:

$$p(E \oplus F) = p(E) \smile p(F) + \beta,$$

where $2\beta = 0$.

For the computation of the Pontrjagin classes of a complex vector bundle the following considerations are useful. Let V be a complex vector space. If we forget that V is a complex vector space and complexify it to obtain $V \otimes_\mathbb{R} \mathbb{C}$, we see that $V \otimes_\mathbb{R} \mathbb{C}$ is, as a complex vector space, isomorphic to $V \oplus \bar{V}$. Namely, $V \otimes_\mathbb{R} \mathbb{C}$ is, as a real vector space, equal to $V \oplus V$ and, with respect to this decomposition, the multiplication by i maps (x, y) to $(-y, x)$.

1. Pontrjagin classes

With this we write down an isomorphism

$$\begin{aligned} V \otimes_{\mathbb{R}} \mathbb{C} = V \oplus V &\longrightarrow V \oplus \bar{V} \\ (x,y) &\longmapsto (x+iy, ix+y). \end{aligned}$$

This extends to vector bundles. For a complex vector bundle E the fibrewise isomorphism above gives an isomorphism:

$$E \otimes_{\mathbb{R}} \mathbb{C} \cong E \oplus \bar{E}.$$

Using the product formula for Chern classes one can express the Pontrjagin classes of a complex vector bundle E in terms of the Chern classes of E. For example:

$$p_1(E) = -c_2(E \oplus \bar{E}) = -(c_1(E) \smile c_1(\bar{E}) + c_2(E) + c_2(\bar{E})) = c_1^2(E) - 2c_2(E).$$

Now, we compute $\langle p_1(E_{k,\ell}), [S^4] \rangle$, where $p: E_{k,\ell} \to S^4$ is the $\mathbb{R}^4$-bundle considered in chapter 17. As for the Euler class one shows that

$$(k, \ell) \longmapsto \langle p_1(E_{k,\ell}), [S^4] \rangle$$

is a homomorphism. Next we observe that $p_1(E_{k,\ell})$ does not depend on the orientation of $E_{k,\ell}$ and, since $E_{k,\ell}$ is isomorphic to $E_{-\ell,-k}$ (reversing orientation), we conclude

$$\langle p_1(E_{k,\ell}), [S^4] \rangle = \langle p_1(E_{-\ell,-k}), [S^4] \rangle.$$

Linearity and this property imply that there is a constant a such that

$$\langle p_1(E_{k,\ell}), [S^4] \rangle = a(k - \ell).$$

To determine a we compute $\langle p_1(E_{0,1}), [S^4] \rangle$. Since, for a fixed element $x \in \mathbb{H}$, the map $y \mapsto y \cdot x$ is $\mathbb{C}$-linear, $E_{0,1}$ is a complex vector bundle. Thus by the formula above:

$$p_1(E_{0,1}) = -2c_2(E_{0,1}) = -2e(E_{0,1}).$$

From $\langle e(E_{k,\ell}), [S^4] \rangle = -k - \ell$ we conclude

$$\langle p_1(E_{0,1}), [S^4] \rangle = 2$$

and thus we have proved:

Proposition 18.1.

$$\langle p_1(E_{k,\ell}), [S^4] \rangle = -2(k - \ell).$$

2. Pontrjagin numbers

To demonstrate the use of characteristic classes we consider the following invariants for closed smooth $4k$-dimensional manifolds M. Let $I := (i_1, i_2, \ldots, i_r)$ be a sequence of natural numbers $0 < i_1 \leq \cdots \leq i_r$ such that $i_1 + \cdots + i_r = k$, i.e., I is a partition of k. Then we define the **Pontrjagin number**

$$p_I(M) := \langle p_{i_1}(TM) \smile \cdots \smile p_{i_r}(TM), [M] \rangle \in \mathbb{Z}.$$

To compute the Pontrjagin numbers in examples we consider the complex projective spaces and look at their tangent bundles. To determine this bundle we consider the following line bundle over $\mathbb{CP}^n$, the **Hopf bundle**. Its total space H is the quotient of $S^{2n+1} \times \mathbb{C}$ under the equivalence relation $(x, z) \sim (\lambda x, \lambda z)$ for some $\lambda \in S^1$. The projection $p : H \to \mathbb{CP}^n$ maps $[(x, z)]$ to $[x]$. The fibre over $[x]$ is equipped with the structure of a 1-dimensional complex vector space by defining $[(x, z)] + [(x, z')] := [(x, z + z')]$. A local trivialization around $[x]$ is given as follows: Let x_i be non-zero and define $U_i := \{[y] \in \mathbb{CP}^n \mid y_i \neq 0\}$. Then a trivialization over U_i is given by the map $p^{-1}(U_i) \to U_i \times \mathbb{C}$ mapping $[(x, z)]$ to $([x], z/x_i)$.

Proposition 18.2. *There is an isomorphism of complex vector bundles*

$$T\mathbb{CP}^n \oplus (\mathbb{CP}^n \times \mathbb{C}) \cong (n+1)H.$$

Proof: We start with the description of $\mathbb{CP}^n$ as

$$\mathbb{C}^{n+1} - \{0\}/\mathbb{C}^* = \mathbb{C}^{n+1} - \{0\}/\sim$$

where $x \sim \lambda x$ for all $\lambda \in \mathbb{C}^*$. Let $\pi : \mathbb{C}^{n+1} - \{0\} \longrightarrow \mathbb{CP}^n$ be the canonical projection. This is a differentiable map. Moreover, if we use complex charts for $\mathbb{CP}^n$, it even is a holomorphic map. Using local coordinates, one checks that for each $x \in \mathbb{C}^{n+1} - \{0\}$ the complex differential $d\pi_x : \mathbb{C}^{n+1} = T_x(\mathbb{C}^{n+1} - \{0\}) \to T_{[x]}\mathbb{CP}^n$ is surjective.

If for some $\lambda \in \mathbb{C}^*$ we consider the map $\mathbb{C}^{n+1} \to \mathbb{C}^{n+1}$ given by multiplication with λ, its complex differential acts on each tangent space $\mathbb{C}^{n+1}$ as multiplication by λ. Thus the differential

$$d\pi : T(\mathbb{C}^{n+1} - \{0\}) \to T\mathbb{CP}^n$$

induces a fibrewise surjective bundle map between two bundles over $\mathbb{CP}^n$

$$[d\pi] : (\mathbb{C}^{n+1} - \{0\}) \times \mathbb{C}^{n+1}/\sim \to T\mathbb{CP}^n$$

where $(x, v) \sim (\lambda x, \lambda v)$. The bundle

$$(\mathbb{C}^{n+1} - \{0\}) \times \mathbb{C}^{n+1}/\sim \to \mathbb{CP}^n$$

2. Pontrjagin numbers

given by projection onto the first factor is $(n+1)H$.

To finish the proof, we have to extend the bundle map $[d\pi]$ to a bundle map
$$(\mathbb{C}^{n+1} - \{0\}) \times \mathbb{C}^{n+1}/\sim \to T\mathbb{C}P^n \oplus (\mathbb{C}P^n \times \mathbb{C})$$
which is fibrewise an isomorphism. This map is given by
$$[x, v] \longmapsto ([d\pi]([x, v]), ([x], \langle v/_{||x||}, x/_{||x||}\rangle))$$
where $\langle v, x \rangle$ is the hermitian scalar product $\Sigma v_i \cdot \bar{x}_i$ and $||x|| = \sqrt{\langle x, x \rangle}$.

Since the kernel of $[d\pi_x]$ consists of all v which are multiples of x, the map is fibrewise injective and thus fibrewise an isomorphism, since both vector spaces have the same dimensions.
q.e.d.

To compute the Pontrjagin classes of the complex projective spaces we have to determine the characteristic classes of H. Since H is a complex line bundle, its first Chern class is equal to $e(H) \in SH^2(\mathbb{C}P^n)$. Since $SH^2(\mathbb{C}P^n)$ is generated by $e(L)$ we know that $e(H) = k \cdot e(L)$ for some k. To determine k it is enough to consider $p : H \to \mathbb{C}P^1$ and to compute $\langle e(H), [\mathbb{C}P^1] \rangle$. For this consider the section $[x] \to [x, x_0]$ which has just one zero at $[x] = [0, 1]$ where it is transverse. One checks that the local orientation at this point is 1. We conclude:
$$\langle e(H), [\mathbb{C}P^1] \rangle = 1$$
and thus
$$c_1(H) = e(H) = -e(L).$$

Now using the relation between the Pontrjagin and Chern classes of a complex bundle above we see that
$$p_1(H) = c_1(H)^2 - 2c_2(H) = e(L)^2,$$
since $c_2(H) = 0$. Thus
$$p(H) = 1 + e(L)^2.$$

With the product formula for Pontrjagin classes and the fact that the cohomology of $\mathbb{C}P^n$ is torsion-free and finitely generated, we conclude from $T\mathbb{C}P^n \oplus (\mathbb{C}P^n \times \mathbb{C}) = (n+1)H$ that $p(T\mathbb{C}P^n) = p((n+1)H)$ and using the product formula again:

Theorem 18.3. *The total Pontrjagin class of the complex projective space* $\mathbb{CP}^n$ *is:*

$$p(T\mathbb{CP}^n) = 1 + p_1(T\mathbb{CP}^n) + \cdots + p_{[n/2]}(T\mathbb{CP}^n) = p(H)^{n+1} = (1 + e(L)^2)^{n+1}$$

or

$$p_k(T\mathbb{CP}^n) = \binom{n+1}{k} \cdot e(L)^{2k}.$$

We use this to compute the following Pontrjagin numbers. We recall that as a consequence of Proposition 11.3 we saw that $e(L) = -x$ and from chapter 11 that $\langle x^n, [\mathbb{CP}^n] \rangle = 1$. Thus $\langle e(L)^{2n}, [\mathbb{CP}^{2n}] \rangle = 1$ and we obtain for example:

$$p_{(1)}(\mathbb{CP}^2) = 3,$$

$$p_{(1,1)}(\mathbb{CP}^4) = 25,$$

$$p_{(2)}(\mathbb{CP}^4) = 10.$$

3. Applications of Pontrjagin numbers to bordism

One of the reasons why Pontrjagin numbers are interesting, is the fact that they are bordism invariants for oriented manifolds. We first note that they are additive under disjoint union and change sign if we pass from M to $-M$ (note that the Pontrjagin classes do not depend on the orientation of a bundle, but the fundamental class does). To see that Pontrjagin numbers are bordism invariants, let W be a compact oriented $(4k+1)$-dimensional smooth manifold with boundary. Using our collar we identify an open neighbourhood of ∂W in W with $\partial W \times [0,1)$. Then $T\overset{\circ}{W}|_{\partial W \times (0,1)} = T\partial W \times ((0,1) \times \mathbb{R})$. Thus from the product formula we conclude: $j^*(p_{i_1}(TW) \smile \cdots \smile p_{i_r}(TW)) = p_{i_1}(T\partial W) \smile \cdots \smile p_{i_r}(T\partial W)$, where j is the inclusion from ∂W to W. From this we see by naturality:

$$\begin{aligned} p_I(\partial W) &= \langle p_{i_1}(T\partial W) \smile \cdots \smile p_{i_r}(T\partial W), [\partial W] \rangle \\ &= \langle p_{i_1}(TW) \smile \cdots \smile p_{i_r}(TW), j_*[\partial W] \rangle = 0, \end{aligned}$$

the latter following since $j_*[\partial W] = 0$ (W is a null bordism!). We summarize:

Theorem 18.4. *The Pontrjagin numbers induce homomorphisms from the oriented bordism group* Ω_{4k} *to* $\mathbb{Z}$:

$$p_I : \Omega_{4k} \longrightarrow \mathbb{Z}.$$

Since $p_n(T\mathbb{CP}^{2n}) = \binom{2n+1}{n} \cdot e(L)^{2n}$, the homomorphism $p_{(n)} : \Omega_{4n} \to \mathbb{Z}$ is non-trivial and we have another proof for the fact we have shown using

3. Applications of Pontrjagin numbers to bordism

the signature, namely that $\Omega_{4k} \neq 0$ for all $k \geq 0$.

The existence of a homomorphism $\Omega_{4k} \to \mathbb{Z}$ for each partition I of k naturally raises the question whether the corresponding elements in $\mathrm{Hom}(\Omega_{4k}, \mathbb{Z})$ are all linearly independent. This is in fact the case and is proved in [**Mi-St**]. In low dimensions one can easily check this by hand. In dimension 4 there is nothing to show. In dimension 8 we consider $\mathbb{CP}^2 \times \mathbb{CP}^2$. The tangent bundle is $T\mathbb{CP}^2 \times T\mathbb{CP}^2$ or $p_1^* T\mathbb{CP}^2 \oplus p_2^* T\mathbb{CP}^2$. Thus by the product formula for the Pontrjagin classes $p_1(T(\mathbb{CP}^2 \times \mathbb{CP}^2)) = p_1^* 3e(L)^2 + p_2^* 3e(L)^2$ and $p_2(T(\mathbb{CP}^2 \times \mathbb{CP}^2)) = p_1^* 3e(L)^2 \smile p_2^* 3e(L)^2 = 9(p_1^* e(L)^2 \smile p_2^* e(L)^2)$ or $9(e(L)^2 \times e(L)^2)$. By definition of the cross product

$$\langle e(L)^2 \times e(L)^2, [\mathbb{CP}^2 \times \mathbb{CP}^2]\rangle = \langle e(L)^2, [\mathbb{CP}^2]\rangle \cdot \langle e(L)^2, [\mathbb{CP}^2]\rangle = 1$$

and so

$$p_{(2)}(\mathbb{CP}^2 \times \mathbb{CP}^2) = 9$$

and, using $p_1(T\mathbb{CP}^2) = 3e(L)^2$ we compute:

$$(p_1(T(\mathbb{CP}^2 \times \mathbb{CP}^2)))^2 = (p_1^* 3e(L)^2 + p_2^* 3e(L)^2)^2$$
$$= 9p_1^* e(L)^4 + 18(p_1^* e(L)^2 \smile p_2^* e(L)^2) + 9p_2^* e(L)^4$$
$$= 18(p_1^* e(L)^2 \smile p_2^* e(L)^2) = 18(e(L)^2 \times e(L)^2).$$

We conclude that

$$p_{(1,1)}(\mathbb{CP}^2 \times \mathbb{CP}^2) = 18.$$

With this information one checks that the matrix

$$\begin{pmatrix} p_{(1,1)}(\mathbb{CP}^4) & p_{(1,1)}(\mathbb{CP}^2 \times \mathbb{CP}^2) \\ p_{(2)}(\mathbb{CP}^4) & p_{(2)}(\mathbb{CP}^2 \times \mathbb{CP}^2) \end{pmatrix} = \begin{pmatrix} 25 & 18 \\ 10 & 9 \end{pmatrix}$$

is invertible and the two homomorphisms on Ω_8 are linearly independent. We summarize:

Theorem 18.5. *The ranks of Ω_4 and Ω_8 satisfy the inequalities*

$$\mathrm{rank}\, \Omega_4 \geq 1$$

and

$$\mathrm{rank}\, \Omega_8 \geq 2.$$

For the first inequality we already have another argument using the signature (Corollary 15.3).

4. Classification of some Milnor manifolds

For a final application of characteristic classes in this section, we return to the Milnor manifolds $M_{k,\ell}$. For dimensional reasons, there is just one Pontrjagin class which might be of some use, namely $p_1(TM_{k,\ell}) \in SH^4(M_{k,\ell}; \mathbb{Z})$. Since this group is torsion except for $k + \ell = 0$ (Proposition 11.4), we only look at $M_{k,-k}$. Since $SH_4(M_{k,-k}) \cong \mathbb{Z}$, there is up to sign a unique generator $[V,g] \in SH_4(M_{k,-k})$. Thus we can obtain a numerical invariant by evaluating $p_1(TM_{k,-k})$ on $[V,g]$ and taking its absolute value:

$$M_{k,-k} \longmapsto |\langle p_1(TM_{k,-k}), [V,g]\rangle|.$$

This is an invariant of the diffeomorphism type of $M_{k,-k}$.

To compute this number, recall that $M_{k,\ell}$ is the sphere bundle of $E_{k,\ell}$. Thus $TM_{k,\ell} \oplus (M_{k,\ell} \times \mathbb{R}) = TD(E_{k,\ell})|_{M_{k,\ell}} = TE_{k,\ell}|_{M_{k,\ell}}$ (for the first identity use a collar of $SE_{k,\ell} = M_{k,\ell}$ in $DE_{k,\ell}$). Let $j : M_{k,\ell} \to E_{k,\ell}$ be the inclusion. Then our invariant is

$$|\langle p_1(TM_{k,-k}), [V,g]\rangle| = |\langle j^*p_1(TE_{k,-k}), [V,g]\rangle|$$
$$= |\langle p_1(TE_{k,-k}), j_*[V,g]\rangle|$$
$$= |\langle p_1(i^*TE_{k,-k}), [S^4]\rangle|.$$

The last equality comes from two facts, namely that the map

$$j_* : H_4(M_{k,-k}) \to H_4(E_{k,-k})$$

is an isomorphism (this follows from a computation of the homology of $E_{k,-k}$ using the Mayer-Vietoris sequence as for $M_{k,-k}$ and comparing these exact sequences) and that the inclusion $i : S^4 \to E_{k,-k}$ given by the zero section induces an isomorphism $SH_4(S^4) \to SH_4(E_{k,-k})$. To compute $p_1(i^*TE_{k,-k}) = p_1(TE_{k,-k}|_{S^4})$, we note that $TE_{k,\ell}|_{S^4} \cong TS^4 \oplus E_{k,\ell}$. The isomorphism is induced by the differential of i from TS^4 to $TE_{k,\ell}$ and by the differential of the inclusion of a fibre $(E_{k,\ell})_x$ to $E_{k,\ell}$ giving a homomorphism $T((E_{k,\ell})_x) = E_{k,\ell} \to TE_{k,\ell}$. With the help of a local trivialization one checks that this bundle map $TS^4 \oplus E_{k,\ell} \to TE_{k,\ell}|_{S^4}$ is fibrewise an isomorphism and thus a bundle isomorphism.

Returning to the Milnor manifolds, since $TS^4 \oplus (S^4 \times \mathbb{R}) = T\mathbb{R}^5|_{S^4} = S^4 \times \mathbb{R}^5$, we note that:

$$|\langle p_1(i^*TE_{k,-k}), [S^4]\rangle| = |\langle p_1(E_{k,-k}), [S^4]\rangle| = 4|k|.$$

Thus $|k|$ is a diffeomorphism invariant of $M_{k,-k}$ as was also the case with L_k. But there is a big difference between the two cases since for L_k we have detected $|k|$ as the order of $SH_1(L_k)$, whereas all $M_{k,-k}$ have the same homology and we have used a more subtle invariant to distinguish them.

Finally, we construct an (orientation-reversing) diffeomorphism from $M_{k,\ell}$ to $M_{-k,-\ell}$ by mapping $D^4 \times S^3$ to $D^4 \times S^3$ via $(x,y) \mapsto (\bar{x},y)$ and $-D^4 \times S^3$ to $-D^4 \times S^3$ via $(x,y) \mapsto (\bar{x},y)$. Thus we conclude:

Theorem 18.6. *Two Milnor manifolds $M_{k,-k}$ and $M_{r,-r}$ are diffeomorphic if and only if $|k| = |r|$.*

5. Exercises

(1) Let E be a complex vector bundle over S^{4k}. Give a formula for the Pontrjagin class $p_k(E)$ in terms of $c_{2k}(E)$.

(2) Let E be a $2k$-dimensional oriented vector bundle. Prove that $p_k(E) = e(E) \smile e(E)$.

(3) Let E be a not necessarily oriented $2k$-dimensional vector bundle. Prove that the class represented by $p_k(E)$ in $\mathbb{Z}/2$-cohomology is equal to $w_{2k}(E) \smile w_{2k}(E)$.

(4) Prove that $\langle p_k(E), [S^{4k}] \rangle$ is even for all vector bundles E over S^{4k}. You can use (or better prove it as an application of Sard's theorem) that an r-dimensional vector bundle over S^n with $r > n$ is isomorphic to $F \oplus (S^n \times \mathbb{R}^{r-n})$ for some n-dimensional vector bundle F.

Chapter 19

Exotic 7-spheres

1. The signature theorem and exotic 7-spheres

At the end of the last section we determined those Milnor manifolds for which $SH_4(M) \cong \mathbb{Z}$. In this chapter we want to look at the other extreme case, namely where all homology groups of $M_{k,\ell}$ except in dimensions 0 and 7 are trivial. By Proposition 11.4 this is equivalent to $k + \ell = \pm 1$. Then homologically $M_{k,\pm 1-k}$ looks like S^7. We are going to prove that it is actually homeomorphic to S^7, a remarkable result by Milnor [**Mi 1**]:

Theorem 19.1. (Milnor): *The Milnor manifolds $M_{k,\pm 1-k}$ are homeomorphic to S^7.*

Although the proof of this result is not related to the main theme of this book we will give it at the end of this chapter for completeness.

This result raises the question whether all manifolds $M_{k,\pm 1-k}$ are diffeomorphic to S^7. We will show that in general this is not the case. We prepare the argument by some considerations concerning bordism groups and the signature.

In chapter 18 we have introduced Pontrjagin numbers, which turned out to be bordism invariants for oriented smooth manifolds. We used them to show that the rank of Ω_4 is at least one and the rank of Ω_8 is at least two. Moreover, the Pontrjagin numbers can be used to show that for all k the products of complex projective spaces $\mathbb{CP}^{2i_1} \times \cdots \times \mathbb{CP}^{2i_r}$ for $i_1 + \cdots + i_r = k$

are linearly independent [**Mi-St**], implying rank $\Omega_{4k} \geq \pi(k)$, the number of partitions of k. In his celebrated paper [**Th 1**] Thom proved that dim $\Omega_{4k} \otimes \mathbb{Q} = \pi(k)$.

Theorem 19.2. (Thom) *The dimension of $\Omega_{4k} \otimes \mathbb{Q}$ is $\pi(k)$ and the products*
$$[\mathbb{CP}^{2i_1} \times \cdots \times \mathbb{CP}^{2i_r}]$$
for $i_1 + \cdots + i_r = k$ form a basis of $\Omega_{4k} \otimes \mathbb{Q}$.

The original proof of this result consists of three steps. The first is a translation of bordism groups into homotopy groups of the so-called Thom space of a certain bundle, the universal bundle over the classifying space for oriented vector bundles. The main ingredient for this so called Pontrjagin-Thom construction is transversality. The second is a computation of the rational cohomology ring. Both steps are explained in the book [**Mi-St**]. The final step is a computation of the rational homotopy groups of this Thom space. Details for this are not given in Milnor-Stasheff, where the reader is referred to the original paper of Serre. An elementary proof based on [**K-K**] is sketched in [**K-L**, p.14 ff].

Now we will apply Thom's result to give a formula for the signature in low dimensions. The key observation here is the bordism invariance of the signature (Theorem 11.6). We recall that the signature induces a homomorphism
$$\tau : \Omega_{4k} \to \mathbb{Z}.$$
Combining this fact with Theorem 19.2 we conclude that the signature can be expressed as a linear combination of Pontrjagin numbers. For example, in dimension 4, where $\Omega_4 \otimes \mathbb{Q} \cong \mathbb{Q}$, the formula can be obtained by comparing $1 = \tau(\mathbb{CP}^2)$ with $\langle p_1(T\mathbb{CP}^2), [\mathbb{CP}^2] \rangle = 3$ and so, for all closed oriented smooth 4-manifolds, one has the formula:
$$\tau(M) = \frac{1}{3} \langle p_1(TM), [M] \rangle.$$

In dimension 8 one knows that there are rational numbers a and b such that
$$\tau(M) = a p_{(1,1)}(M) + b p_{(2)}(M) = a \langle p_1(TM)^2, [M] \rangle + b \langle p_2(TM), [M] \rangle.$$
We have computed the Pontrjagin numbers of $\mathbb{CP}^2 \times \mathbb{CP}^2$ and $\mathbb{CP}^4$. We know already that $\tau(\mathbb{CP}^4) = 1$ and one checks that also $\tau(\mathbb{CP}^2 \times \mathbb{CP}^2) = 1$ (or uses the product formula for the signature). Comparing the values of the signature and the Pontrjagin numbers for these two manifolds one concludes:

1. The signature theorem and exotic 7-spheres

Theorem 19.3. (Hirzebruch) *For a closed oriented smooth 8-dimensional manifold M one has*

$$\tau(M) = \frac{1}{45}(7\langle p_2(TM), [M]\rangle - \langle p_1(TM)^2, [M]\rangle).$$

Proof: We only have to check the formula for $\mathbb{CP}^4$ and for $\mathbb{CP}^2 \times \mathbb{CP}^2$. The values for the Pontrjagin numbers were computed at the end of chapter 18 and with this the reader can verify the formula.
q.e.d.

The two formulas above are special cases of Hirzebruch's famous **Signature Theorem**, which gives a corresponding formula in all dimensions (see [**Hir**] or [**Mi-St**]).

One of the most spectacular applications of Theorem 19.3 was Milnor's discovery of exotic spheres. Milnor shows that in general $M_{k,1-k}$ is not diffeomorphic to S^7. His argument is the following: Suppose there is a diffeomorphism $f : M_{k,1-k} \to S^7$. Since $M_{k,1-k}$ is the boundary of the disk bundle $DE_{k,1-k}$, we can then form the closed smooth manifold

$$N := DE_{k,1-k} \cup_f D^8.$$

We extend the orientation of $DE_{k,1-k}$ to N (which can be done, since the disk has an orientation-reversing diffeomorphism) and compute its signature. The inclusion induces an isomorphism $j^* : SH^4(N) \cong SH^4(\overset{\circ}{DE}_{k,1-k}) \cong SH^4(S^4) \cong \mathbb{Z}$. We will show that the signature of N is -1 by constructing a class with negative self-intersection number. To do this we consider the cohomology class $j^*([S^4, v]) \in SH^4(\overset{\circ}{DE}_{k,1-k})$, where v is the zero-section. We also consider $v^*(j^*([S^4, v])) \in SH^4(S^4)$. This is equal to the Euler class of $E_{k,1-k}$. By definition the self-intersection $S_N([S^4, v], [S^4, v])$ is equal to $\langle e(E_{k,1-k}), [S^4]\rangle$. We have computed this number in Proposition 16.7 and conclude:

$$S_N([S^4, v], [S^4, v]) = -k - (1 - k) = -1.$$

Thus:

$$\tau(N) = -1.$$

Now we use the Signature Theorem to compute $\tau(N)$ in terms of the characteristic numbers $\langle p_1(TN)^2, [N]\rangle$ and $\langle p_2(TN), [N]\rangle$. Since the map $v^* : SH^4(N) \to SH^4(S^4)$ is an isomorphism, we conclude that $p_1(TN) =$

$(v^*)^{-1}(p_1(TN|_{S^4}))$. But $v^*TN \cong TS^4 \oplus E_{k,1-k}$ and then it follows from the Whitney formula and Proposition 18.1 that

$$\langle v^*(p_1(TN)), [S^4]\rangle = \langle p_1(E_{k,1-k}), [S^4]\rangle = -2(2k-1).$$

Comparing this information with the Kronecker product $\langle v^*([S^4, v]), [S^4]\rangle = -1$ we conclude:

$$p_1(TN) = 2(2k-1)[S^4, v].$$

Using

$$S_N([S^4, v], [S^4, v]) = -k - (1-k) = -1$$

we have:

$$\langle p_1^2(TN), [N]\rangle = -4(2k-1)^2.$$

Now we feed this information into the Signature Theorem 19.3:

$$-1 = \tau(N) = \frac{1}{45}(7\langle p_2(TN), [N]\rangle + 4(2k-1)^2).$$

Since $\langle p_2(TN), [N]\rangle \in \mathbb{Z}$, we obtain the congruence

$$45 + 4(2k-1)^2 \equiv 0 \mod 7$$

if $M_{k,1-k}$ is diffeomorphic to S^7. Taking $k=2$ we obtain a contradiction and so have proved:

Theorem 19.4. (Milnor) *$M_{2,-1}$ is homeomorphic, but not diffeomorphic, to S^7.*

This was the first example of a so-called exotic smooth structure on a manifold, i.e., a second smooth structure which is not diffeomorphic to the given one.

We give another application of the signature formula. Given a topological manifold M of dimension $2k$ one can ask whether there is a complex structure on M, i.e., an atlas of charts in $\mathbb{C}^k$ whose coordinate changes are holomorphic functions. We suppose now that M is closed and connected. A necessary condition is that M admits a non-trivial class in $SH_{2k}(M)$. One can introduce the concept of orientation for topological manifolds and show that a connected closed n-dimensional manifold is orientable if and only if a non-trivial class in $SH_n(M)$ exists. Thus the necessary condition above amounts to a topological version of orientability. If $k=1$ it is a classical fact, that all orientable surfaces admit a complex structure. As another application of the signature formula we show:

Theorem 19.5. *S^4 admits no complex structure.*

Proof: If S^4 is equipped with a complex structure, the tangent bundle is a complex vector bundle. For a complex vector bundle E we can compute the first Pontrjagin class using the formula from chapter 18:
$$p_1(E) = -2c_2(E).$$
Thus
$$p_1(TS^4) = -2c_2(TS^4) = -2e(TS^4),$$
since $c_1(TS^4) = 0$. Now we use the fact from the Remark after Corollary 12.2 that $\langle e(TM), [M] \rangle = e(M)$ (following from the Poincaré-Hopf Theorem for vector fields) and conclude:
$$\langle e(TS^4), [S^4] \rangle = e(S^4) = 2.$$
Next we note that $\tau(S^4) = 0$ and so we obtain a contradiction from the signature formula:
$$0 = \tau(S^4) = 1/3 \langle p_1(TS^4), [S^4] \rangle = -4/3.$$

q.e.d.

One actually can show that S^{2k} has no complex structure for $k \neq 1, 3$. It is a famous open problem whether S^6 has a complex structure.

2. The Milnor spheres are homeomorphic to the 7-sphere

We finish this chapter with the proof of Theorem 19.1. It is based on an elementary but fundamental argument in Morse theory.

Lemma 19.6. *Let W be a compact smooth manifold with $\partial W = M_0 \sqcup M_1$. If there is a smooth function*
$$f : W \to [0, 1]$$
without critical points and $f(M_0) = 0$ and $f(M_1) = 1$, then W is diffeomorphic to $M_0 \times [0, 1]$.

Proof: We try to give a self-contained presentation, for background information see [**Mi 3**]. Choose a smooth Riemannian metric g on TW (for example, embed W smoothly into an Euclidean space and restrict the Euclidean metric to each fibre of the tangent bundle). Consider the so-called **normed gradient vector field** of f which is defined by mapping $x \in M$ to the tangent vector $s(x) \in T_x M$ such that

i) $df_x s(x) = 1 \in \mathbb{R} = T_{f(x)}\mathbb{R}$,

ii) $\langle s(x), v \rangle_{g(x)} = 0$ for all v with $df_x(v) = 0$.

This is a well defined function since the dimension of ker df_x is dim $M-1$ and $df_x|_{\ker df_x^\perp}$ is an isomorphism (the orthogonal complement is taken with respect to g_x). Since W, f and g are smooth, this is a smooth vector field on W.

Now, we consider the ordinary differential equation for a given point $x \in W$:
$$\dot{\varphi}(t) = s(\varphi(t)) \text{ and } \varphi(0) = x,$$
where φ is a smooth function (a path) from an interval to W and as usual we abbreviate the differential of a path φ at the time t by $\dot{\varphi}(t)$.

The existence and uniqueness result for ordinary differential equations says that locally (using a chart to translate everything into $\mathbb{R}^m$) there is a unique solution called an integral curve. Furthermore, the solution depends smoothly on the initial point x and t.

Now, for each $x \in M_0$ we consider a maximal interval for which one has a solution φ_x with initial value x. Then
$$df(\dot{\varphi}_x(t)) = df(s(\varphi_x(t))) = 1.$$
Thus
$$f(\varphi_x(t)) = t + c$$
for some $c \in \mathbb{R}$. Since $\varphi_x(0) = x$, we conclude $c = 0$ and so $f(\varphi_x(t)) = t$.

Since W is compact, the interval is maximal and since $f(\varphi_x(t)) = t$, the interval has to be $[0,1]$. As φ_x depends smoothly on x and t we obtain a smooth function
$$\begin{aligned} \psi : M_0 \times [0,1] &\to W \\ (x,t) &\mapsto \varphi_x(t). \end{aligned}$$

This function is a diffeomorphism since it has an inverse. For this, consider for $y \in W$ the integral curve of the differential equation:
$$\dot{\eta}_y(t) = -s(\eta_y(t)) \text{ and } \eta_y(0) = y$$

2. The Milnor spheres are homeomorphic to the 7-sphere

(we use the negative gradient field to "travel" backwards). As above, we see that
$$f(\eta_y(t)) = f(y) - t.$$
The integral curve $\eta_{\varphi_x(t)}$ joins $\varphi_x(t)$ with x and is the time inverse of the integral curve φ_x. With this information, we can write down the inverse:
$$\psi^{-1}(y) = (\eta_y(f(y)), f(y)).$$

q.e.d.

Proof of Theorem 19.1 after Milnor: For simplicity we only consider the case $M_{k,1-k}$; the other case follows similarly. With Lemma 19.6 we will give the proof by constructing two disjoint embeddings D^7_+ and D^7_- in $M_{k,1-k}$ and constructing a smooth function
$$f : M_{k,1-k} - (\mathring{D}^7_+ + \mathring{D}^7_-) \to [0,1]$$
without critical points. Then by Lemma 19.6 there is a diffeomorphism
$$\varphi : S^6_+ \times [0,1] \longrightarrow M_{k,1-k} - (\mathring{D}^7_+ + \mathring{D}^7_-)$$
with $\varphi(x, 0) = x$ for all $x \in S^6_+$.

From this we construct a homeomorphism from $M_{k,1-k}$ to $S^7 = D^7_+ \cup D^7_-$ as follows. We map
$$x \in D^7_+ \text{ to } x \in D^7_+ \subset S^7,$$
$$\varphi(x,t) \text{ to } (1-t/2)\cdot x \in D^7_- \text{ for } x \in S^6_+ \text{ and } t \in [0,1],$$
$$x \in D^7_- \text{ to } x/2 \in D^7_- \subset S^7.$$
The reader should check that this map is well defined, continuous and bijective. Thus it is a homeomorphism.

Continuing with the proof, we note that
$$M_{k,\ell} = \mathbb{H} \times S^3 \cup_{f_{k,\ell}} -\mathbb{H} \times S^3$$
where $f_{k,\ell} : \mathbb{H} - \{0\} \times S^3 \to -\mathbb{H} - \{0\} \times S^3$ maps
$$(x,y) \mapsto (x/\|x\|^2, x^k y x^\ell / \|x\|^{(k+\ell)}).$$

We have used this description since it gives $M_{k,\ell}$ as a smooth manifold. Now we consider the smooth functions
$$\begin{aligned} g: \mathbb{H} \times S^3 &\longrightarrow \mathbb{R} \\ (x,y) &\longmapsto \frac{y_1}{\sqrt{1+\|x\|^2}} \end{aligned}$$

and
$$h: -\mathbb{H} \times S^3 \longrightarrow \mathbb{R}$$
$$(x,y) \longmapsto \frac{(x \cdot y^{-1})_1}{\sqrt{1+||x \cdot y^{-1}||^2}}$$
where $(\ )_1$ denotes the first component.

If $\ell = 1 - k$, the two functions are compatible with the gluing function $f_{k,1-k}$ and thus
$$g \cup h : M_{k,1-k} \longrightarrow \mathbb{R}$$
is a smooth function.

What are the singular points of g and h? The function h has no singular points but g has singular points $(0, 1)$ and $(0, -1)$, where $1 = (1, 0, 0, 0) \in S^3$. Thus, 1 and -1 are the only singular values of $g \cup h$.

Since $\pm 1/2$ are regular values, we can decompose the manifold $M_{k,1-k}$ as $(g \cup h)^{-1}(-\infty, -\frac{1}{2}] \cup (g \cup h)^{-1}[-\frac{1}{2}, \frac{1}{2}]$ and $(g \cup h)^{-1}[\frac{1}{2}, \infty) =: D_+ \cup W \cup D_-$. We identify $D_\pm$ with D^7 as follows: $D_+ = (g \cup h)^{-1}(-\infty, -\frac{1}{2}] = \{(x, y) \in \mathbb{H} \times S^3 \mid y_1 \leq -\frac{1}{2}\sqrt{1 + ||x||^2}\}$ using the fact that $y \in S^3$ and so $y_1^2 + y_2^2 + y_3^2 + y_4^2 = 1$. From this we conclude that
$$D_+ = \{(x, (y_2, y_3, y_4)) \mid 4(y_2^2 + y_3^2 + y_4^2) + ||x||^2 \leq 3\}$$
and thus D_+ is diffeomorphic to D^7. Similarly one shows that D_- is diffeomorphic to D^7. Since $g \cup h|_W$ has no critical points, we may now apply Lemma 19.6.
q.e.d.

3. Exercises

(1) Prove that there is no complex structure on $-\mathbb{CP}^2$, i.e., no complex structure whose underlying oriented manifold has the opposite orientation of $\mathbb{CP}^2$.

Chapter 20

Relation to ordinary singular (co)homology

1. $SH_k(X)$ is isomorphic to $H_k(X;\mathbb{Z})$ for CW-complexes

This chapter has a different character since we use several concepts and results which are not covered in this book. In particular we assume familiarity with ordinary singular homology and cohomology.

Eilenberg and Steenrod showed that if a functor on the category of finite CW-complexes X (actually they consider finite polyhedra, but up to homotopy equivalence this is the same as finite CW-complexes) fulfills certain homology axioms, then there is a unique natural isomorphism between this homology theory and ordinary singular homology $H_k(X)$, which for a point is the identity [**E-S**]. Their axioms are equivalent to our axioms, if in addition the homology groups of a point are $\mathbb{Z}$ in degree 0 and 0 otherwise. Thus for finite CW-complexes X there is a unique natural isomorphism (which for a point is the identity)

$$\sigma : SH_k(X) \to H_k(X).$$

Since $SH_k(X)$ is compactly supported one can extend σ to a natural transformation for arbitrary CW-complexes. Namely, if X is a CW-complex and $[\mathbf{S}, g]$ is an element of $SH_k(X)$, the image of $\mathbf{S}$ under g is compact. Thus there is a finite subcomplex Y in X such that $g(\mathbf{S}) \subset Y$. Let $i : Y \to X$ be the inclusion, then we consider $i_*(\sigma([\mathbf{S}, g]) \in H_k(X)$, where we consider

[**S**, g] as element of $SH_k(Y)$. It is easy to see that this gives a well defined natural transformation

$$\sigma : SH_k(X) \to H_k(X)$$

for arbitrary CW-complexes X. We use the fact that if (**T**, h) is a bordism, then $g(\mathbf{T})$ is contained in some other finite subcomplex Z with $Y \subset Z$.

Theorem 20.1. *The natural transformation*

$$\sigma : SH_k(X) \to H_k(X)$$

is an isomorphism for all CW-complexes X and all k.

This natural transformation commutes with the $\times$-product.

More generally it is enough to require that X is homotopy equivalent to a CW-complex. All smooth manifolds are homotopy equivalent to CW-complexes [**Mi 3**] and so theorem 20.1 holds for all smooth manifolds.

Proof: We know this already for finite CW-complexes. The argument for arbitrary CW-complexes uses the same idea as the construction of the generalization of σ. Namely if X is an arbitrary CW-complex and $x \in H_k(X)$ is a homology class then there exists a finite subcomplex Y such that $x \in \text{im}(H_k(Y) \to H_k(X))$. From this we conclude using the result for finite CW-complexes that x is in the image of $\sigma : SH_k(X) \to H_k(X)$. Similarly, if $x \in SH_k(X)$ maps to zero under σ, we find a finite CW-complex $Z \subset X$ such that $x \in \text{im}(SH_k(Z) \to SH_k(X))$ since $SH_k(X)$ has compact supports. Thus we can assume that $x \in SH_k(Z)$. Since $H_k(X)$ has compact supports there is a finite CW complex $T \subset X$ such that $Z \subset T$ and $\sigma(x)$ maps to zero in $H_k(T)$. From the result for finite CW-complexes we conclude $\sigma(x) = 0$ in $H_k(T)$ and so $x = 0$.

To show that the natural transformation commutes with the $\times$-product we use a description of ordinary singular homology using bordism of regular oriented **parametrized** stratifolds (p-stratifolds) instead of arbitrary stratifolds. The same arguments as for bordism groups of general stratifolds show that this is a homology theory. However, there is a difference, namely by a Mayer-Vietoris argument one shows that every closed oriented parametrized p-stratifold **S** has a fundamental class in ordinary homology $[\mathbf{S}] \in H_n(\mathbf{S})$ and one obtains a natural transformation from the bordism group bases on p-stratifolds to ordinary homology by mapping (**S**, f) to $f_*([\mathbf{S}]) \in H_n(X)$. By the comparison theorem for homology theories this is an isomorphism. It is an easy argument with the Künneth formula for ordinary homology to show that this natural transformation preserves the $\times$-product. The forgetful map

2. An example where $SH_k(X)$ and $H_k(X)$ are different

(forgetting the parametrization) gives another natural transformation from homology based on parametrized stratifolds to $SH_k(X)$ which preserves the ×-product. Since the natural transformations commute for CW-complexes (by the fact that for a point they are the identity using the uniqueness result mentioned above from [**E-S**]) this shows that the natural transformation above commutes with the ×-product.
q.e.d.

Remark: A similar argument gives a natural isomorphism
$$\sigma_n : SH_k(X; \mathbb{Z}/2) \to H_k(X; \mathbb{Z}/2)$$
for all CW-complexes X.

2. An example where $SH_k(X)$ and $H_k(X)$ are different

We denote the oriented surface of genus g by F_g. For $g = 1$ we obtain the torus $F_1 = T$ and F_g is the connected sum of g copies of the torus.

We consider the following subspace of $\mathbb{R}^3$ given by an infinite connected sum of tori as in the following picture, where the point on the right side is removed. We call this an infinite sum of tori. This is a non-compact smooth submanifold of $\mathbb{R}^3$ denoted by F_∞. The space in the picture is the one-point compactification of F_∞. This is a compact subspace of $\mathbb{R}^3$.

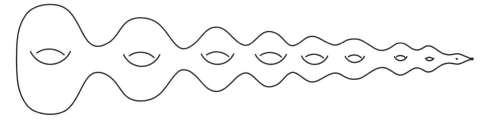

As in example (2) in chapter 2, section 3 (page 21), we make F_∞^+ a 2-dimensional stratifold denoted **S** by the algebra **C** consisting of continuous functions which are constant near the additional point and smooth on F_∞. Obviously, this stratifold is regular and oriented. Thus we can consider the fundamental class
$$[\mathbf{S}] = [\mathbf{S}, \mathrm{id}] \in SH_2(\mathbf{S}).$$

This class has the following property. Let $p_g : \mathbf{S} \to F_g$ be the projection onto F_g (we map all tori added to F_g to obtain F_∞ to a point). Then
$$(p_g)_*([\mathbf{S}]) = [F_g]$$

(why?). In particular $(p_g)_*([\mathbf{S}])$ is non-trivial for all g.

But there is no class α in $H_2(\mathbf{S})$ such that $p_{g*}(\alpha)$ is non-trivial for all g. The reason is that for each topological space X and each class α in $H_2(X)$ there is a map $f : F \to X$, where F is a closed oriented surface, such that $\alpha = f_*([F])$. This follows from [**C-F**] using the Atiyah-Hirzebruch spectral sequence. Now we suppose that we can find $f : F \to \mathbf{S}$ such that $(p_g)_*(\alpha) \neq 0$ in $H_2(F_g)$ for all g. But this is impossible since the degree of fp_g is non-zero and there is no map $F \to F_g$ with degree non-zero if the genus of F is smaller than g. The reason is that if the degree is non-trivial then the induced map $H_1(F_g) \to H_1(F)$ is injective (as follows from the regularity of the intersection form over $\mathbb{Q}$, Corollary 14.6).

We summarize these considerations:

Theorem 20.2. *The homology theories $SH_k(X)$ and $H_k(X)$ are not equivalent for general topological spaces.*

3. $SH^k(M)$ is isomorphic to ordinary singular cohomology

We also want to identify our cohomology groups $SH^k(M)$ constructed via stratifolds with the singular cohomology groups $H^k(M)$.

So far we only have defined integral cohomology groups for oriented manifolds. In the exercises of chapter 13 (page 129) we extended the definition of cohomology groups and induced maps to arbitrary manifolds. We want to compare this cohomology theory with ordinary singular cohomology on smooth manifolds.

We use a characterization of singular cohomology on smooth manifolds from [**K-S**]. The main result of this paper says that we only have to check the following condition for such a cohomology theory h for which the cohomology groups of a point are $\mathbb{Z}$ in degree 0 and 0 otherwise.
For $i = 1, 2 \ldots$ let M_i be a sequence of smooth manifolds. Then

$$h^k(\bigsqcup M_i) \cong \prod_i h^k(M_i),$$

where the isomorphism from $h^k(\bigsqcup M_i)$ to the direct product is induced by the inclusions.

3. $SH^k(M)$ is isomorphic to ordinary singular cohomology

Since this condition holds for our cohomology theory there is a unique natural isomorphism θ from $SH^k(M)$ to $H^k(M)$ inducing the identity on $SH^0(\text{pt})$ [**K-S**].

In the same paper the multiplicative structure is also characterized. The standard cup product (or equivalently $\times$-product) on $H^*(M)$ is characterized in [**K-S**] by the property, that if for a closed oriented manifold M the class i_M is the Kronecker dual to the fundamental class, then for $S^k \times S^n$ we have:
$$i_{S^k} \times i_{S^n} = i_{S^k \times S^n}.$$
One can reformulate this condition without referring to the Kronecker product by characterizing i_M as the unique class which for each oriented chart $\varphi: U \to \mathbb{R}^m$ corresponds to the generator of $H^m(\mathbb{R}^m, \mathbb{R}^m - 0) \cong H^{m-1}(S^{m-1}) = \mathbb{Z}$ under the maps $H^m(\mathbb{R}^m, \mathbb{R}^m - 0) \to H^m(U, U - x) \cong H^m(M, M - x) \to H^m(M)$. Using this characterization of i_M for stratifold cohomology one can for the product in $SH^*(S^k \times S^n)$ check the condition above.

Summarizing we obtain:

Theorem 20.3. *There is a unique natural isomorphism θ from the cohomology groups of manifolds constructed in this book via stratifolds to ordinary singular cohomology, commuting with the $\times$-products and inducing the identity on cohomology in degree 0.*

Since the natural transformation θ respects the cup product we obtain a geometric interpretation of the intersection form on ordinary singular cohomology. Let M be a closed smooth oriented manifold of dimension m. Since θ respects cup products we conclude:

Corollary 20.4. *For a closed smooth oriented m-dimensional manifold M and cohomology classes $x \in H^k(M)$ and $y \in H^{m-k}(M)$ we have the identity:*
$$\langle x \smile y, [M] \rangle = [\mathbf{S}_x, g_x] \pitchfork [\mathbf{S}_y, g_y],$$
where $[S_x, g_x] := \theta(x)$ and $[S_y, g_y] := \theta(y)$ are cohomology classes in $SH^k(M)$ and $SH^{m-k}(M)$ corresponding to x and y via θ and $\pitchfork$ means the transverse intersection.

Thus the traditional geometric interpretation of the intersection form for those cohomology classes on a closed oriented smooth manifold, where the Poincaré duals are represented by maps from closed oriented smooth manifolds to M, as a transverse intersection makes sense for arbitrary cohomology classes.

The natural isomorphism between the (co)homology groups defined in this book and ordinary singular cohomology allows us, for CW-complexes, to translate results from one of the worlds to the other. Above we have made use of this by interpreting the intersection form on singular cohomology geometrically. The geometric feature is one of the strengths of our approach to (co)homology. There are other aspects of (co)homology which are easier and more natural in ordinary singular (co)homology, in particular those which allow an application of homological algebra. This is demonstrated by the general Künneth Theorem or by the various universal coefficient theorems. It is useful to have both interpretations of (co)homology available so that one can choose in which world one wants to work depending on the questions one is interested in.

4. Exercises

(1) Let $\pi : \widetilde{X} \to X$ be a covering space with a constant finite number of points in each fibre. Let $\mathbf{S}$ be a compact oriented regular stratifold of dimension n and $f : \mathbf{S} \to X$ be a continuous map. One defines the pull-back $f^*(\widetilde{X}) := \{(s,x) \in \mathbf{S} \times \widetilde{X} \,|\, f(s) = \pi(x)\}$. Show that the projection to the first factor $p : f^*(\widetilde{X}) \to \mathbf{S}$ is a covering map and so $f^*(\widetilde{X})$ is a compact oriented regular stratifold of dimension n where the orientation of $f^*(\widetilde{X})$ is the one such that the projection map will be orientation-preserving. The projection to the second factor gives a map to $\widetilde{X}$ and thus an element in $SH_n(\widetilde{X})$. Show that this induces a well defined map $SH_n(X) \to SH_n(\widetilde{X})$ denoted by $\pi^!$ (called the **transfer map**) and that the composition $SH_n(X) \to SH_n(\widetilde{X}) \to SH_n(X)$ is multiplication by the number of points in the fibre.

Appendix A

Constructions of stratifolds

1. The product of two stratifolds

Now we show that $(\mathbf{S} \times \mathbf{S}', \mathbf{C}(\mathbf{S} \times \mathbf{S}'))$ as defined in chapter 2 is a stratifold. It is clear that $\mathbf{S} \times \mathbf{S}'$ is a locally compact Hausdorff space with countable basis. We have to show that $\mathbf{C}(\mathbf{S} \times \mathbf{S}')$ is an algebra. Let f and g be in $\mathbf{C}(\mathbf{S} \times \mathbf{S}')$, $x \in \mathbf{S}^i$ and $y \in (\mathbf{S}')^j$. Using local retractions one sees that $f + g$ and fg are in $\mathbf{C}(\mathbf{S} \times \mathbf{S}')$. Obviously, the constant maps are in $\mathbf{C}(\mathbf{S} \times \mathbf{S}')$. Since we characterize $\mathbf{C}(\mathbf{S} \times \mathbf{S}')$ by local conditions it is locally detectable. Also the condition in the definition of a differential space is obvious.

Next we show that restriction gives an isomorphism of germs at $(x, y) \in \mathbf{S}^i \times (\mathbf{S}')^j$:

$$\mathbf{C}(\mathbf{S} \times \mathbf{S}')_{(x,y)} \xrightarrow{\cong} C^\infty(\mathbf{S}^i \times (\mathbf{S}')^j)_{(x,y)}.$$

To see that this map is surjective, we consider $f \in C^\infty(\mathbf{S}^i \times (\mathbf{S}')^j)$ and choose for x a local retraction $r : U \to V$ near x of $\mathbf{S}$ and for y a local retraction $r' : U' \to V'$ near y of $\mathbf{S}'$. Let ρ be a smooth function on $\mathbf{S}$ with support $\rho \subset U$ which is constant 1 near x and ρ' a corresponding smooth function on $\mathbf{S}'$ with support $\rho' \subset U'$ which is constant 1 near y. Then $\rho(z)\rho'(z')f(r(z), r'(z'))$ (which we extend by 0 to the complement of $U \times U'$) is in $\mathbf{C}(\mathbf{S} \times \mathbf{S}')$. (To see this we only have to check for $(z, z') \in U \times U'$ that there are local retractions q near z and q' near z' such that $f(rq(t), r'q'(t')) = f(r(t), r'(t'))$. But since r is a morphism, we can choose q such that $rq(t) = r(t)$ and similarly $r'q'(t') = r'(t')$ implying the statement.) Thus we have found a germ at (x, y) which maps to f under restriction.

To see that the map is injective, we note that if $f \in \mathbf{C}(\mathbf{S} \times \mathbf{S}')$ maps to zero in $C^\infty(\mathbf{S}^i \times (\mathbf{S}')^j)_{(x,y)}$ it vanishes in an open neighbourhood of (x, y) in $\mathbf{S}^i \times (\mathbf{S}')^j$ and since there are retractions near x and y such that f commutes with them, f is zero in some open neighbourhood of (x, y) in $\mathbf{S} \times \mathbf{S}'$.

Having shown that $\mathbf{C}(\mathbf{S} \times \mathbf{S}')_{(x,y)} \xrightarrow{\cong} C^\infty(\mathbf{S}^i \times (\mathbf{S}')^j)_{(x,y)}$ is an isomorphism, we conclude that $T_{(x,y)}(\mathbf{S} \times \mathbf{S}') \cong T_{(x,y)}(\mathbf{S}^i \times (\mathbf{S}')^j)$, and so the induced stratification on $\mathbf{S} \times \mathbf{S}'$ is given by $\bigsqcup_{i+j=k} \mathbf{S}^i \times (\mathbf{S}')^j$. Now condition 1 of a stratifold also follows from the isomorphism of germs, condition 2 is obvious and condition 3 follows from the product $\rho\rho'$ of appropriate bump functions of $\mathbf{S}$ and $\mathbf{S}'$.

Thus $(\mathbf{S} \times \mathbf{S}', \mathbf{C}(\mathbf{S} \times \mathbf{S}'))$ is a stratifold.

2. Gluing along part of the boundary

In the proof of the Mayer-Vietoris sequence we will also need gluing along part of the boundary. If one glues naively then corners or cusps occur (see the figure at the top of the following page). In a natural way the corners or cusps can be removed or better smoothed. The central tool for this smoothing is given by collars. The constructions will depend on the choice of a collar, not just on the corresponding germ. However, up to bordism, these choices are irrelevant.

Now we return to gluing along part of the boundary. Consider two c-stratifolds W_1 and W_2 and suppose that ∂W_1 is obtained by gluing two c-stratifolds Z and Y_1 over the common boundary $\partial Z = \partial Y_1 = N$ (assuming that Z and Y_1 have collars φ_Z and φ_{Y_1}): $\partial W_1 = Z \cup_N Y_1$. Similarly, we assume that $\partial W_2 = Z \cup_N Y_2$ (using collars φ_Z and φ_{Y_2}) and that W_1 and W_2 have collars η_1 and η_2. Then we want to make $W_1 \cup_{\overset{\circ}{Z}} W_2$ a c-stratifold with boundary $Y_1 \cup_N Y_2$. We define $W_1 \cup_{\overset{\circ}{Z}} W_2$ as $W_1 \cup_Z W_2 - Y_1 \cup_N Y_2$. But this space is equal to $W_1 - Y_1 \cup_{\overset{\circ}{Z}} W_2 - Y_2$, gluing of two c-stratifolds along the full boundary $\overset{\circ}{Z}$, which is a stratifold by the considerations above. If we add the boundary $Y_1 \cup_N Y_2$ naively and use the given collars, we obtain "cusps" along N.

2. Gluing along part of the boundary 193

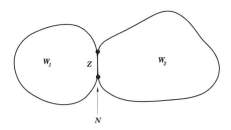

To smooth along N we first combine φ_Z and φ_{Y_1} to an isomorphism $\varphi_1 : N \times (-1,1) \to \partial W_1$ onto its image, where $\varphi_1(x,t) := \varphi_Z(x,t)$ for $t \geq 0$ and $\varphi_1(x,t) := \varphi_{Y_1}(x,-t)$ for $t \leq 0$. $\varphi_1|_{N\times\{0\}}$ is the identity map. Similarly, we combine φ_Z and φ_{Y_2} to $\varphi_2 : N \times (-1,1) \to \partial W_1$ and note that $\varphi_2|_{N\times[0,1)} = \varphi_1|_{N\times[0,1)}$. We denote by $\alpha_1 : N \times (-1,1) \times [0,1) \to W_1$ the map $(x,s,t) \mapsto \eta_1(\varphi_1(x,s),t)$. We denote the image by U_1. This map is an isomorphism away from the boundary. Similarly, we define $\alpha_2 : N \times (-1,1) \times [0,1) \to U_2$. The union $U_1 \cup U_2 := U_N$ is an open neighbourhood of N in $W_1 \cup_Z W_2$.

Now we pass in $\mathbb{R}^2$ to polar coordinates (r,φ) and choose a smooth monotone map $\rho : \mathbb{R}_{\geq 0} \to (0,1]$, which is equal to $\frac{1}{2}$ for $r \leq \frac{1}{3}$ and equal to 1 for $r \geq \frac{2}{3}$ (it is important to fix this map for the future and use the same map to make the constructions unique). Then consider the map β_1 from $(-1,1) \times [0,1) \subset \{(r,\varphi)|r \geq 0, 0 \leq \varphi \leq \pi\}$ to $\mathbb{R}^2$ mapping (r,φ) to $(r,\rho(r)\cdot\varphi)$ and similarly β_2 mapping (r,φ) to $(r,-\rho(r)\cdot\varphi)$. The images of $(-1,1) \times [0,1)$ in cartesian coordinates look roughly like

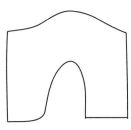

and

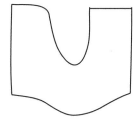

Identifying $\beta_1([0,1) \times \{0\})$ with $\beta_2([0,1) \times \{0\})$ gives a smooth c-manifold G looking as follows

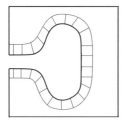

where the collar is indicated in the figure. We obtain a homeomorphism $\Phi : U_N \longrightarrow N \times G$ mapping $\alpha_1(x, s, t)$ to $(x, \beta_1(s, t))$ and $\alpha_2(x, s, t)$ to $(x, \beta_2(s, t))$. The map Φ is an isomorphism of stratifolds outside N. By construction the collar induced from $N \times G$ via Φ and the collars of W_1 along $\partial W_1 - \operatorname{im} \varphi_{Y_1}$ and of W_2 along $\partial W_2 - \operatorname{im} \varphi_{Y_2}$ fit together to give a collar on $W_1 \cup_Z W_2$ finishing the proof of:

Proposition A.1. *For $i = 1, 2$ let W_i be c-stratifolds such that ∂W_i is obtained by gluing two c-stratifolds Z and Y_i over the common boundary $\partial Z = \partial Y_i = N$:*

$$\partial W_i = Z \cup_N Y_i.$$

Choose representatives of the germs of collars for Y_i and Z.

Then there is a c-stratifold $W_1 \cup_Z W_2$ extending the stratifold structures on $W_i - (Z \cup \operatorname{im} \varphi_{Y_i})$. The boundary of $W_1 \cup_Z W_2$ is $Y_1 \cup_N Y_2$.

It should be noted that the construction of the collar of $W_1 \cup_Z W_2$ depends on the choice of representatives of the collars of W_i, Y_i and Z. For our application in the proof of the Mayer-Vietoris sequence it is important to observe that the collar was constructed in such a way that, away from the neighbourhood of the union of the collars of N in Y_i and Z, it is the original collar of W_1 and W_2.

3. Proof of Proposition 4.1

We conclude this appendix by proving that for a space X the isomorphism classes of pairs $(\mathbf{S}, g)$, where $\mathbf{S}$ is an m-dimensional stratifold, and $g : \mathbf{S} \to X$ is a continuous map, form a set.

Proof of Proposition 4.1: For this we first note that the diffeomorphism classes of manifolds form a set. This follows since a manifold is diffeomorphic to one obtained by taking a countable union of open subsets of $\mathbb{R}^m$ (the domains of a countable atlas) and identifying them according to an appropriate equivalence relation. Since the countable sum of copies of $\mathbb{R}^m$ forms a set, the set of subsets of a set forms a set, and the possible equivalence relations on these sets form a set, the diffeomorphism classes of m-dimensional

3. Proof of Proposition 4.1

manifolds are a subset of the set of all sets obtained from a countable disjoint union of subsets of $\mathbb{R}^m$ by some equivalence relation.

Next we note that a stratifold is obtained from a disjoint union of manifolds, the strata, by introducing a topology (a collection of certain subsets) and a certain algebra. The possible topologies as well as the possible algebras are a set. Thus the isomorphism classes of stratifolds are a set. Finally for a fixed stratifold $\mathbf{S}$ and space X the maps from $\mathbf{S}$ to X are a set, and so we conclude that the isomorphism classes of pairs $(\mathbf{S}, g)$, where $\mathbf{S}$ is an m-dimensional stratifold, and $g : \mathbf{S} \to X$ is a continuous map, form a set.
q.e.d.

Appendix B

The detailed proof of the Mayer-Vietoris sequence

The following lemma is the main tool for completing the proof of the Mayer-Vietoris sequence along the lines explained in §5. It is also useful in other contexts. Roughly it says that up to bordism we can separate a regular stratifold **S** by an open cylinder over some regular stratifold **P**. Such an embedding is called a **bicollar**, i.e., an isomorphism $g : \mathbf{P} \times (-\epsilon, \epsilon) \to V$, where V is an open subset of **S**. The most naive idea would be to "replace" **P** by $\mathbf{P} \times (-\epsilon, \epsilon)$, so that as a set we change **S** into $(\mathbf{S} - \mathbf{P}) \cup (\mathbf{P} \times (-\epsilon, \epsilon))$. The proof of the following lemma makes this rigorous.

Lemma B.1. *Let $\mathbf{T}$ be a regular c-stratifold. Let $\rho : \mathbf{T} \to \mathbb{R}$ be a continuous function such that $\rho|_{\overset{\circ}{\mathbf{T}}}$ is smooth. Let 0 be a regular value of $\rho|_{\overset{\circ}{\mathbf{T}}}$ and suppose that $\rho^{-1}(0) \subset \overset{\circ}{\mathbf{T}}$ and that there is an open neighbourhood of 0 in $\mathbb{R}$ consisting only of regular values of $\rho|_{\overset{\circ}{\mathbf{T}}}$.*

Then there exists a regular c-stratifold $\mathbf{T}'$ and a continuous map $f : \mathbf{T}' \to \mathbf{T}$ with $\partial \mathbf{T}' = \partial \mathbf{T}$, $f|_{\partial \mathbf{T}'} = \mathrm{id}$ such that f commutes with appropriate representatives of the collars of $\mathbf{T}'$ and $\mathbf{T}$. Furthermore, there is an $\epsilon > 0$ such that $\rho^{-1}(0) \times (-\epsilon, \epsilon)$ is contained in $\mathbf{T}'$ as an open subset and a continuous map $\rho' : \mathbf{T}' \to \mathbb{R}$ whose restriction to the interior is smooth and whose restriction to $\rho^{-1}(0) \times (-\epsilon, \epsilon)$ is the projection to $(-\epsilon, \epsilon)$. The restriction of f to $\rho^{-1}(0) \times (-\epsilon, \epsilon)$ is the projection onto $\rho^{-1}(0)$. In addition

$(\rho')^{-1}(-\infty, -\epsilon) \subset \rho^{-1}(-\infty, 0)$ and $(\rho')^{-1}(\epsilon, \infty) \subset \rho^{-1}(0, \infty)$.

If $\partial \mathbf{T} = \varnothing$, then $(\mathbf{T}, \mathrm{id})$ and $(\mathbf{T}', f)$ are bordant.

Proof: Choose $\delta > 0$ such that such that $(-\delta, \delta)$ consists only of regular values of ρ.

Consider a monotone smooth map $\mu : \mathbb{R} \to \mathbb{R}$ which is the identity for $|t| > \delta/2$ and 0 for $|t| < \delta/4$.

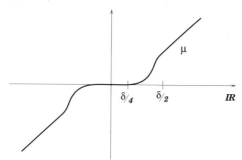

Then $\eta : \mathbf{T} \times \mathbb{R} \to \mathbb{R}$ mapping $(x, t) \mapsto \rho(x) - \mu(t)$ has 0 as a regular value. Namely, for those (x, t) mapping to 0 with $|t| < \delta$ we have $|\rho(x)| < \delta$ and thus (x, t) is a regular point of η, and for those (x, t) mapping to 0 with $|t| > \delta/2$ we have $\mu(t) = t$ and again (x, t) is a regular point. Thus $\mathbf{T}' := \eta^{-1}(0)$ is, by Proposition 4.2, a regular c-stratifold (the collar is discussed below) containing $V := \rho^{-1}(0) \times (-\delta/4, \delta/4)$. Setting $\epsilon = \delta/4$ we have constructed the desired subset in $\mathbf{T}'$.

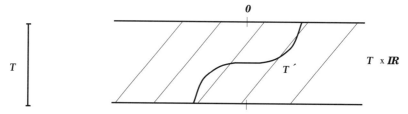

To relate $\mathbf{T}'$ to $\mathbf{T}$, consider the map $f : \mathbf{T}' \to \mathbf{T}$ given by the restriction of the projection onto $\mathbf{T}$ in $\mathbf{T} \times \mathbb{R}$. This is an isomorphism outside $\rho^{-1}(0) \times (-\delta/2, \delta/2)$. In particular we can identify the boundaries via this isomorphism: $\partial \mathbf{T}' = \partial \mathbf{T}$. Similarly we use this isomorphism to induce a collar on $\mathbf{T}'$ from a small collar of $\mathbf{T}$ and so the c-structure on $\mathbf{T}$ makes $\mathbf{T}'$ a regular c-stratifold. Finally we define ρ' by the projection onto $\mathbb{R}$. The desired properties are obvious and this finishes the proof of the first statement.

If $\partial \mathbf{T} = \varnothing$, we want to construct a bordism between $(\mathbf{T}, \mathrm{id})$ and $(\mathbf{T}', f)$. For this, choose a smooth map $\zeta : I \to \mathbb{R}$ which is 0 near 0 and 1 near

B. The detailed proof of the Mayer-Vietoris sequence 199

1. Then consider the smooth map $\mathbf{T} \times \mathbb{R} \times I \to \mathbb{R}$ mapping $(x,t,s) \mapsto \rho(x) - (\zeta(s)\mu(t) + (1-\zeta(s))t)$. This map again has 0 as regular value and the preimage of 0 is the required bordism $\mathbf{Q}$. By construction and Proposition 4.2 $\mathbf{Q}$ is a regular c-stratifold. The projection from $\mathbf{Q}$ to $\mathbf{T}$ is a map $r : \mathbf{Q} \to \mathbf{T}$, whose restriction to $\mathbf{T}$ is the identity on $\mathbf{T}$ and whose restriction to $\mathbf{T}'$ is f. Thus $(\mathbf{Q}, r)$ is a bordism between $(\mathbf{T}, \mathrm{id})$ and $(\mathbf{T}', f)$.
q.e.d.

Now we apply this lemma to the proof of Proposition 5.1 and the detailed proof of Theorem 5.2, the Mayer-Vietoris sequence.

Proofs of Proposition 5.1 and Theorem 5.2: We begin with the proof of Proposition 5.1. For $[\mathbf{S}, g] \in SH_m(X)$ we consider (as before Proposition 5.1) the closed subsets $A := g^{-1}(X - V)$ and $B := g^{-1}(X - U)$. Using a partition of unity we construct a smooth function $\rho : \mathbf{S} \to \mathbb{R}$ and choose a regular value s such that $\rho^{-1}(s) \subset \mathbf{S} - (A \cup B)$ and $A \subset \rho^{-1}(s, \infty)$ and $B \subset \rho^{-1}(-\infty, s)$. After composition with an appropriate translation we can assume $s = 0$. Since $\mathbf{S}$ is compact, by Proposition 4.3 the regular values of ρ form an open set in $\mathbb{R}$.

Thus we can apply Lemma B.1 and we consider $\mathbf{S}'$, f and ρ'. Then $(\mathbf{S}, g)$ is bordant to $(\mathbf{S}', gf)$ (since $(\mathbf{S}', f)$ is bordant to $(\mathbf{S}, \mathrm{id})$) and 0 is a regular value of ρ'. By construction, $\rho^{-1}(0) \times (-\epsilon, \epsilon)$ is contained in $\mathbf{S}'$ as open neighbourhood of $\mathbf{P} := (\rho')^{-1}(0) = \rho^{-1}(0)$, in other words we have a bicollar of $\mathbf{P}$. Furthermore by construction gf is equal to g on $\mathbf{P} = \rho^{-1}(0)$, in particular, $gf(\mathbf{P})$ is contained in $U \cap V$. In Proposition 5.1 we defined $d([\mathbf{S}, g])$ as $[\rho^{-1}(0), g|_{\rho^{-1}(0)}]$ and the considerations so far imply that this definition is the same if we pass from $(\mathbf{S}, g)$ to the bordant pair $(\mathbf{S}', gf)$ and define $d([\mathbf{S}', gf])$ as $[\mathbf{P}, gf|_{\mathbf{P}}]$: this situation has the advantage that $\mathbf{P}$ has a bicollar.

To show that d is well defined it is enough to show that if $(\mathbf{S}', gf)$ is the boundary of $(\mathbf{T}, F)$, then $[\mathbf{P}, g|_{\mathbf{P}}]$ is zero in $SH_{k-1}(U \cap V)$. Here $\mathbf{T}$ is a c-stratifold with boundary $\mathbf{S}'$. In particular we can take as $\mathbf{T}$ the cylinder over $\mathbf{S}$ and see that d does not depend on the choice of the separating function or the regular value. We choose a representative of the germ of collars $\mathbf{c}$ of $\mathbf{T}$. Define $A_{\mathbf{T}} := F^{-1}(A)$ and $B_{\mathbf{T}} := F^{-1}(B)$ and construct a smooth function $\eta : \mathbf{T} \to \mathbb{R}$ with the following properties:
1) There is a $\mu > 0$ such that the restriction of η to $\mathbf{P} \times (-\mu, \mu)$ is the projection to $(-\mu, \mu)$,
2) $\eta(\mathbf{c}(x,t)) = \eta(x)$,

3) there is an $\delta > 0$ such that $F(\eta^{-1}(-\delta, \delta)) \subset U \cap V$.
The construction of such a map η is easy using a partition of unity since $\mathbf{P}$ has a bicollar in $\mathbf{S}'$.

By Sard's theorem there is a t with $|t| < \min\{\delta, \mu\}$ which is a regular value of η. Since the restriction of η to $\mathbf{P} \times (-\mu, \mu)$ is the projection to $(-\mu, \mu)$ we conclude that t is also a regular value of $\eta|_{\mathbf{S}'}$. By condition 2) we guarantee that $\mathbf{Q} := \eta^{-1}(t)$ is a c-stratifold with boundary $\mathbf{P} \times \{t\}$. By condition 3) we know that $F(\mathbf{Q}) \subset U \cap V$ and so we see that $[\mathbf{P} \times \{t\}, F|_{\mathbf{P} \times \{t\}}]$ is zero in $SH_{k-1}(U \cap V)$. On the other hand, obviously $[\mathbf{P} \times \{t\}, F|_{\mathbf{P} \times \{t\}}]$ is bordant to $[\mathbf{P}, g|_{\mathbf{P}}]$. This finishes the proof of Proposition 5.1.

Now we proceed to the proof of Theorem 5.2. We first show that d commutes with induced maps. The reason is the following. Let X' be a space with decomposition $X' = U' \cup V'$ and $h : X \to X'$ a continuous map respecting the decomposition. Then if we consider $(\mathbf{S}, hf)$ instead of $(\mathbf{S}, f)$, one can take the same separating function ρ in the definition of d and so $d'([\mathbf{S}, hf]) = [\rho^{-1}(s), hf|_{\rho^{-1}(s)}] = h_*([\rho^{-1}(s), f|_{\rho^{-1}(s)}]) = h_*(d([\mathbf{S}, f]))$.

Now we begin the proof of the exactness by examining
$$SH_n(U \cap V; \mathbb{Z}/2) \to SH_n(U; \mathbb{Z}/2) \oplus SH_n(V; \mathbb{Z}/2) \to SH_n(U \cup V; \mathbb{Z}/2).$$

Since $j_U i_U = i : U \cap V \to U \cup V$, the inclusion map, and also $j_V i_V = i$, the difference of the composition of the two maps is zero. To show the reverse inclusion, consider $[\mathbf{S}, f] \in SH_n(U; \mathbb{Z}/2)$ and $[\mathbf{S}', f'] \in SH_n(V; \mathbb{Z}/2)$ with $(j_U)_*([\mathbf{S}, f]) = (j_V)_*([\mathbf{S}', f'])$. Let $(\mathbf{T}, g)$ be a bordism between $[\mathbf{S}, f]$ and $[\mathbf{S}', f']$, where $g : \mathbf{T} \to U \cup V$. Similarly, as in the proof that d is well defined, we consider the closed disjoint subsets $A_{\mathbf{T}} := \mathbf{S} \cup g^{-1}(X - V)$ and $B_{\mathbf{T}} := \mathbf{S}' \cup g^{-1}(X - U)$. Using a partition of unity we construct a smooth function $\rho : \mathbf{T} \to \mathbb{R}$ with $\rho(A) = -1$ and $\rho(B) = 1$ and choose a regular value s such that $\rho^{-1}(s) \subset \overset{\circ}{\mathbf{T}} - (A_{\mathbf{T}} \cup B_{\mathbf{T}})$. After composition with an appropriate translation we can assume $s = 0$. Since $\mathbf{T}$ is compact, by Proposition 4.3, the regular values of ρ form an open set in $\mathbb{R}$. Applying Lemma B.1 we can assume after replacing $\mathbf{T}$ by $\mathbf{T}'$ that $\rho^{-1}(s)$ has a bicollar φ. Then $[\rho^{-1}(s), g|_{\rho^{-1}(s)}] \in SH_n(U \cap V)$ and — as explained in §3 — $(\rho^{-1}[s, \infty), g|_{\rho^{-1}[s, \infty)})$ is a bordism between $(\mathbf{S}, f)$ and $(\rho^{-1}(s), g|_{\rho^{-1}(s)})$ in U.

B. The detailed proof of the Mayer-Vietoris sequence

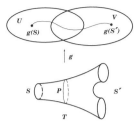

Similarly we obtain by $(\rho^{-1}(-\infty, s]), g|_{\rho^{-1}(-\infty,s]})$ a bordism between $(\mathbf{S}', f')$ and $(\rho^{-1}(s), g|_{\rho^{-1}(s)})$ in V. Thus

$$((i_U)_*([\rho^{-1}(s), g|_{\rho^{-1}(s)}]), (i_V)_*([\rho^{-1}(s), g|_{\rho^{-1}(s)}])) = ([\mathbf{S}, f], [\mathbf{S}', f']).$$

Next we consider the exactness of

$$SH_n(U \cup V; \mathbb{Z}/2) \xrightarrow{d} SH_{n-1}(U \cap V; \mathbb{Z}/2) \to SH_{n-1}(U; \mathbb{Z}/2) \oplus SH_{n-1}(V; \mathbb{Z}/2).$$

By construction of the boundary operator the composition of the two maps is zero. Namely $(\rho^{-1}[s, \infty), f|_{\rho^{-1}[s,\infty)})$ is a null cobordism of $d([\mathbf{S}, f])$ in U and $(\rho^{-1}(-\infty, s], f|_{\rho^{-1}(-\infty,s]})$ is a zero-bordism of $d([\mathbf{S}, f])$ in V. Here we again apply Lemma B.1 and assume that $\rho^{-1}(s)$ has a bicollar.

To show the reverse inclusion, start with $[\mathbf{P}, r] \in SH_{n-1}(U \cap V; \mathbb{Z}/2)$ and suppose $(i_U)_*([\mathbf{P}, r]) = 0$ and $(i_V)_*([\mathbf{P}, r]) = 0$. Let $(\mathbf{T}_1, g_1)$ be a zero bordism of $(i_U)_*([\mathbf{P}, r])$ and $(\mathbf{T}_2, g_2)$ be a zero bordism of $(i_V)_*([\mathbf{P}, r])$. Then we consider $\mathbf{T}_1 \cup_\mathbf{P} \mathbf{T}_2$. Since $g_1|_\mathbf{P} = g_2|_\mathbf{P} = r$, we can, as in the proof of the transitivity of the bordism relation, extend r to $\mathbf{T}_1 \cup_\mathbf{P} \mathbf{T}_2$ using g_1 and g_2 and denote this map by $g_1 \cup g_2$. Thus $[\mathbf{T}_1 \cup_\mathbf{P} \mathbf{T}_2, g_1 \cup g_2] \in SH_n(U \cup V; \mathbb{Z}/2)$. By the construction of the boundary operator we have $d([\mathbf{T}_1 \cup_\mathbf{P} \mathbf{T}_2, g_1 \cup g_2]) = [\mathbf{P}, r]$. Using the bicollar one constructs a separating function which near $\mathbf{P}$ is the projection from $\mathbf{P} \times (-\epsilon, \epsilon)$ to the second factor.

Finally, we prove exactness of

$$SH_n(U; \mathbb{Z}/2) \oplus SH_n(V; \mathbb{Z}/2) \to SH_n(U \cup V; \mathbb{Z}/2) \xrightarrow{d} SH_{n-1}(U \cap V; \mathbb{Z}/2).$$

The composition of the two maps is obviously zero. Now, consider $[\mathbf{S}, f] \in SH_n(U \cup V; \mathbb{Z}/2)$ with $d([\mathbf{S}, f]) = 0$. Consider ρ, s and $\mathbf{P}$ as in the definition of the boundary map d and assume by Lemma B.1 that $\rho^{-1}(s)$ has a bicollar. We put $\mathbf{S}_+ := \rho^{-1}[s, \infty)$ and $\mathbf{S}_- := \rho^{-1}(-\infty, s]$. Then $\mathbf{S} = \mathbf{S}_+ \cup_\mathbf{P} \mathbf{S}_-$. If $d([\mathbf{S}, f]) = [\mathbf{P}, f|_\mathbf{P}] = 0$ in $SH_{n-1}(U \cap V; \mathbb{Z}/2)$ there is $\mathbf{Z}$ with $\partial \mathbf{Z} = \mathbf{P}$ and an extension of $f|_\mathbf{P}$ to $r : \mathbf{Z} \to U \cap V$. We glue to obtain $\mathbf{S}_+ \cup_\mathbf{P} \mathbf{Z}$ and $\mathbf{S}_- \cup_\mathbf{P} \mathbf{Z}$. Since $f|_\mathbf{P} = r|_\mathbf{P}$ the maps $f|_{\mathbf{S}_+}$ and r give a continuous map $f_+ : \mathbf{S}_+ \cup_\mathbf{P} \mathbf{Z} \to U$ and similarly we obtain $f_- : \mathbf{S}_- \cup_\mathbf{P} \mathbf{Z} \to V$. We are finished if $(j_U)_*([S_+ \cup_P Z, f_+]) - (j_V)_*([S_- \cup_P Z, f_-]) = [S, f]$. For this we have to find a bordism (T, g) such that $\partial T = S_+ \cup Z + S_- \cup Z + S$ (recall that $-[P, r] = [P, r]$ for all elements in $SH_n(Y; \mathbb{Z}/2)$) and g extends the given three maps on the pieces.

This bordism is given as $T := ((S_+ \cup_P Z) \times [0, 1]) \cup_Z ((S_- \cup_P Z) \times [1, 2])$ with $\partial T = (S_+ \cup_P Z) \times \{0\} + (S_- \cup_P Z) \times \{2\} + S_+ \cup_P S_-$. Here we apply Lemma A.1 to smooth the corners or cusps. This finishes the proof of Theorem 5.2.
q.e.d.

Finally we discuss the modification needed to prove the Mayer-Vietoris sequence in cohomology. Everything works with appropriate obvious modifications as for homology except where we argue that the regular values of the separating map ρ form an open set. This used the fact that the stratifold on which ρ is defined is compact, which is not the case for regular stratifolds representing cohomology classes. The separating function can be chosen as we wish, and we show now that we can always find a separating function ρ and a regular value, which is an interior point of the set of regular values.

Let $g : S \to M$ be a proper smooth map and C and D be disjoint closed subsets of M. We choose a smooth map $\rho : M \to \mathbb{R}$ which on C is 1 and on D is -1. We select a regular value s of ρg. The set of singular points of ρg is closed by Proposition 4.3, and since a proper map on a locally compact space is closed [Sch, p. 72], the image of the singular points of ρg under g is a closed subset F of M.

Now we consider a bicollar $\varphi : U \to M - F$ of $\rho^{-1}(s)$, where $U = \{(x, t) \in \rho^{-1}(s) \times \mathbb{R} \mid |t| < \delta(x)\}$ for some continuous map $\delta : \rho^{-1}(s) \to \mathbb{R}_{>0}$. We can choose φ in such a way that $\rho\varphi(x, t) = t$. Now we "expand" this bicollar by choosing a diffeomorphism from U to $\rho^{-1}(s) \times (-1/2, 1/2)$ mapping (x, t) to $(x, \eta(x, t))$, where $\eta(x, \cdot)$ is a diffeomorphism for each $x \in \rho^{-1}(s)$. Using this, it is easy to find a new separating function ρ', such that $\rho'\varphi(x, t) = t$ and $\rho'^{-1}(-1/2, 1/2) = U$. By construction the interval $(-1/2, 1/2)$ consists only of regular values of $\rho' f$.

We apply this in the proof of the Mayer-Vietoris sequence for cohomology as follows. Let U and V be open subsets of $M = U \cup V$. We consider the closed subsets $C := M - U$ and $D := M - V$. Then we construct ρ' as above and note that $\rho' g$ is a separating function of $A := g^{-1}(C)$ and $B := g^{-1}(D)$, and s is a regular value which is an interior point of the set of

B. The detailed proof of the Mayer-Vietoris sequence

regular values. With this the definition of the coboundary operator works as explained in chapter 12.

Now we explain why the Mayer-Vietoris sequence is exact. We shall explain each argument in figuress, with a brief description followed by a sequence of four figures presented on the immediately following page.

We begin with the exactness of

$$SH^{k-1}(U \cap V) \to SH^k(U \cup V) \to SH^k(U) \oplus SH^k(V).$$

Let $\alpha \in SH^k(U \cup V)$ (figure A) such that it maps to zero, i.e., there are stratifolds with boundary and proper maps extending the map representing α after restricting to U and V respectively. We abbreviate these extensions by β and γ and write $\partial \beta = j_U^*(\alpha)$ and $\partial \gamma = j_V^*(\alpha)$ (figure B). Now we restrict β and γ to the intersection $U \cap V$ and glue them (respecting the orientations) along the common boundary to obtain $\zeta := (-\gamma|_{U \cap V}) \cup \beta|_{U \cap V} \in SH^{k-1}(U \cap V)$ (figure C). Using a separating function ρ we determine the image of ζ under the coboundary operator: $\delta(\zeta)$. Finally we have to show that $\delta(\zeta)$ is bordant to α. For this we consider $\eta := \beta|_{\rho^{-1}(-\infty,s]} \cup (-\gamma|\rho^{-1}[s,\infty))$, which gives such a bordism (figure D).

B. The detailed proof of the Mayer-Vietoris sequence

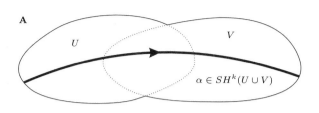

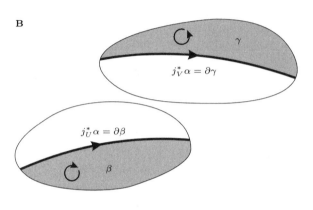

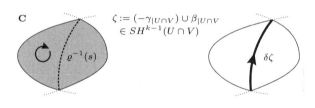

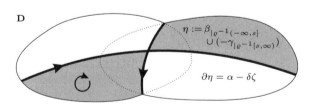

B. The detailed proof of the Mayer-Vietoris sequence

Now we consider the exactness of the sequence (see next page for figures)
$$SH^k(U \cup V) \to SH^k(U) \oplus SH^k(V) \to SH^k(U \cap V).$$

For this we consider $\alpha \in SH^k(U)$ and $\beta \in SH^k(V)$ (figure A) such that (α, β) maps to zero in $SH^k(U \cap V)$. This means there is γ, a stratifold with boundary together with a proper map to $U \cap V$, such that $\partial \gamma = i_U^*(\alpha) - i_V^*(\beta)$ (figure B). Next we choose a separating function ρ as indicated in figure B. Using ρ we consider $\zeta := \alpha_{|\varrho^{-1}(-\infty,s]} \cup (-\delta\gamma) \cup \beta_{|\varrho^{-1}[s,\infty)} \in SH^k(U \cup V)$ (figure C). Finally we have to construct a bordism between $j_U^*(\zeta)$, and α resp. $j_V^*(\zeta)$ and β. This is given by the equations $j_U^*\zeta + \partial(\gamma_{|\varrho^{-1}[s,\infty)}) = \alpha$ and $j_V^*\zeta - \partial(\gamma_{|\varrho^{-1}(-\infty,s]}) = \beta$ (figure D).

206   B. The detailed proof of the Mayer-Vietoris sequence

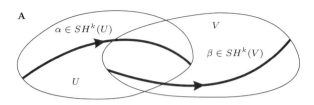

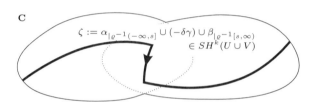

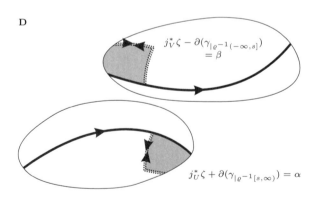

B. The detailed proof of the Mayer-Vietoris sequence

Finally we consider the exactness of the sequence (see next page for figures)
$$SH^k(U) \oplus SH^k(V) \to SH^k(U \cap V) \to SH^{k+1}(U \cup V).$$
Let α be in $SH^k(U \cap V)$ (and ρ a separating function) such that $\delta\alpha = 0$ (figure A). This means that there is a stratifold β with boundary $\delta(\alpha)$ and a proper map extending the given map (figure B). From this we construct the classes $\zeta_1 := \alpha_{|\rho^{-1}(-\infty,s]} \cup \beta_{|U} \in SH^k(U)$ and $\zeta_2 := (-\alpha_{|\rho^{-1}[s,\infty)}) \cup (-\beta_{|V}) \in SH^k(V)$ (figure C). Finally we note that $i_U^*(\zeta_1) - i_V^*(\zeta_2) = \alpha$ (figure D).

208   B. The detailed proof of the Mayer-Vietoris sequence

A

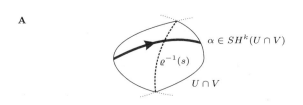

B

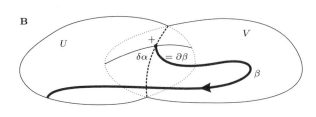

C

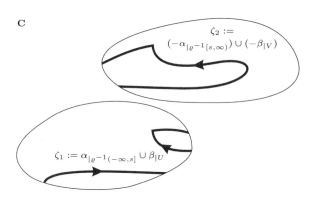

D

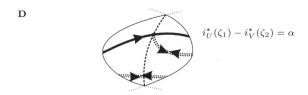

Appendix C

The tensor product

We want to describe an important construction in linear algebra, the tensor product. Let R be a commutative ring with unit, for example $\mathbb{Z}$ or a field. The tensor product assigns to two R-modules another R-module. The slogan is: bilinearity is transferred to linearity. Consider a bilinear map $f : V \times W \to P$ between R-modules. Then we will construct another R-module denoted $V \otimes_R W$ together with a canonical map $V \times W \to V \otimes_R W$ such that f induces a map from $V \otimes_R W \to P$ whose precomposition with the canonical map is f.

Since we are particularly interested in the case of $R = \mathbb{Z}$ we note that a $\mathbb{Z}$-module is the same as an abelian group. If A is an abelian group we make it a $\mathbb{Z}$-module by defining (for $n \geq 0$) $n \cdot a := a + \cdots + a$, where the sum is taken over n summands, and for $n < 0$ we define $n \cdot a := -(-n \cdot a)$.

We begin with the definition of $V \otimes_R W$. This is an R-module generated by all pairs (v, w) with $v \in V$ and $w \in W$. One denotes the corresponding generators by $v \otimes w$ and calls them **pure tensors**. The fact that these will be the generators means that we will obtain a surjective map

$$\bigoplus_{(v,w) \in V \times W} (v, w) \cdot R \longrightarrow V \otimes_R W$$

mapping (v, w) to $v \otimes w$. In order to finish the definition of $V \otimes_R W$ we only need to define the kernel K of this map. We describe the generators of the kernel, which are:

$$(rv, w) - (v, rw) \text{ for all } v \in V, w \in W, r \in R \text{ and}$$
$$(rv, w) - (v, w)r \text{ for all } v \in V, w \in W, r \in R \text{ and}$$
$$(v, w) + (v', w) - (v + v', w), \text{ respectively,}$$
$$(v, w) + (v, w') - (v, w + w') \text{ for all } v, v' \in V, w, w' \in W.$$

Let K be the submodule generated by these elements. Then we define the **tensor product**

$$V \otimes_R W := \left(\bigoplus_{(v,w) \in V \times W} (v, w) \cdot R \right) \Big/ K.$$

Remark: *The following rules are translations of the relations and very useful for working with tensor products:*

$$r \cdot (v \otimes w) = (r \cdot v) \otimes w = v \otimes (r \cdot w)$$
$$v \otimes w + v' \otimes w = (v + v') \otimes w$$
$$v \otimes w + v \otimes w' = v \otimes (w + w').$$

These rules imply that the following **canonical map** is well defined and bilinear:
$$\begin{array}{rcl} V \times W & \longrightarrow & V \otimes_R W \\ (v, w) & \longmapsto & v \otimes w. \end{array}$$

Let $f : V \times W \to P$ be bilinear. Then f induces a linear map
$$\begin{array}{rcl} f : V \otimes_R W & \longrightarrow & P \\ v \otimes w & \longmapsto & f(v, w). \end{array}$$

This map is well defined since $(rv) \otimes w - v \otimes (rw) \mapsto f(rv, w) - f(v, rw) = rf(v, w) - rf(v, w) = 0$ and $v \otimes w + v' \otimes w - (v + v') \otimes w \mapsto f(v, w) + f(v', w) - f(v + v, w) = 0$, respectively, $v \otimes w + v \otimes w' - v \otimes (w + w') \mapsto 0$.

In turn, if we have a linear map from $V \otimes_R W$ to P, the composition of the canonical map with this map is a bilinear map from $V \times W$ to P. Thus as indicated above we have seen the fundamental fact:

The linear maps from $V \otimes_R W$ to P correspond isomorphically to the bilinear maps from $V \times W$ to P.

What is $(V \oplus V') \otimes_R W$? The reader should convince himself that the following maps are bilinear:

C. The tensor product

$$\begin{aligned}(V \oplus V') \times W &\longrightarrow (V \otimes_R W) \oplus (V' \otimes_R W)\\ ((v,v'),w) &\longmapsto (v \otimes w, v' \otimes w)\end{aligned}$$

and

$$\begin{aligned}V \times W &\longrightarrow (V \oplus V') \otimes_R W \quad \text{and} \quad V' \times W \longrightarrow (V \oplus V') \otimes_R W\\ (v,w) &\longmapsto (v,0) \otimes w \qquad\qquad\qquad\qquad (v',w) \longmapsto (0,v') \otimes w.\end{aligned}$$

These maps induce homomorphisms

$$\begin{aligned}(V \oplus V') \otimes_R W &\longrightarrow (V \otimes_R W) \oplus (V' \otimes_R W)\\ (v,v') \otimes w &\longmapsto (v \otimes w, v' \otimes w)\end{aligned}$$

and

$$\begin{aligned}(V \otimes_R W) \oplus (V' \otimes_R W) &\longrightarrow (V \oplus V') \otimes_R W\\ ((v \otimes w_1),(v' \otimes w_2)) &\longmapsto (v,0) \otimes w_1 + (0,v') \otimes w_2\end{aligned}$$

and these are inverse to each other. Thus we have shown:

Proposition C.1. $(V \oplus V') \otimes_R W \xrightarrow{\cong} (V \otimes_R W) \oplus (V' \otimes_R W)$.

It follows that

$$R^n \otimes_R R^m = (R^{n-1} \oplus R) \otimes_R R^m \cong (R^{n-1} \otimes_R R^m) \oplus (R \otimes_R R^m)$$
$$\cong (R^{n-1} \otimes_R R^m) \oplus (R \otimes_R [R \oplus \cdots \oplus R]) = (R^{n-1} \otimes_R R^m) \oplus R^m.$$

Thus dim $R^n \otimes_R R^m = n \cdot m$ and

$$\begin{aligned}R^n \otimes_R R^m &\cong R^{n \cdot m} \cong M(n,m)\\ e_i \otimes e_j &\longmapsto e_{i,j}\end{aligned}$$

where $e_{i,j}$ denotes the $n \times m$ matrix whose coefficients are 0 except at the place (i,j) where it is 1.

Example:
$$\begin{aligned}R \otimes_R M &\cong M\\ r \otimes x &\mapsto r \cdot x.\end{aligned}$$
The inverse is $x \mapsto 1 \otimes x$.

If $R = \mathbb{Z}$, a $\mathbb{Z}$-module is the same as an abelian group. For abelian groups A and B we write $A \otimes B$ instead of $A \otimes_{\mathbb{Z}} B$.

We want to determine $\mathbb{Z}/n \otimes \mathbb{Z}/m$. We prepare this by some general considerations. Let $f : A \to B$ and $g : C \to D$ be homomorphisms of R-modules. They induce a homomorphism

$$\begin{aligned}f \otimes g: \quad A \otimes_R C &\to B \otimes_R D\\ a \otimes c &\mapsto f(a) \otimes g(c),\end{aligned}$$

called the **tensor product** of f and g.

If we have an exact sequence of R-modules
$$\cdots \to A_{k+1} \to A_k \to A_{k-1} \to \cdots$$
and a fixed R-module P, we can tensor all A_k with P and tensor all maps in the exact sequence with id on P, and obtain a new sequence of maps
$$\cdots \to A_{k+1} \otimes_R P \to A_k \otimes_R P \to A_{k-1} \otimes_R P \to \cdots$$
called the induced sequence and ask if this is again exact. This is, in general, not the case and this is one of the starting points of homological algebra, which systematically investigates the failure of exactness. Here we only study a very special case.

Proposition C.2. *Let*
$$0 \to A \to B \to C \to 0$$
be a short exact sequence of R-modules. Then the induced sequence
$$A \otimes_R P \to B \otimes_R P \to C \otimes_R P \to 0$$
is again exact. In general the map $A \otimes_R P \to B \otimes_R P$ is not injective.

Proof: Denote the map from $A \to B$ by f and the map from B to C by g. Obviously $(g \otimes \text{id})(f \otimes \text{id})$ is zero. Thus $g \otimes \text{id}$ induces a homomorphism $B \otimes_R P /_{(f \otimes \text{id})(A \otimes_R P)} \to C \otimes_R P$. We have to show that this is an isomorphism. We give an inverse by defining a bilinear map $C \times P$ to $B \otimes_R P /_{(f \otimes \text{id})(A \otimes_R P)}$ by assigning to (c, p) an element $[b \otimes p]$, where $g(b) = c$. The exactness of the original sequence shows that this induces a well defined homomorphism from $C \otimes_R P$ to $B \otimes_R P /_{(f \otimes \text{id})(A \otimes_R P)}$ and that it is an inverse of $B \otimes_R P /_{(f \otimes \text{id})(A \otimes_R P)} \to C \otimes_R P$.

The last statement follows from the next example.
q.e.d.

As an application we compute $\mathbb{Z}/n \otimes \mathbb{Z}/m$. For this consider the exact sequence
$$0 \to \mathbb{Z} \to \mathbb{Z} \to \mathbb{Z}/n \to 0$$
where the first map is multiplication by n, and tensor it with $\mathbb{Z}/m$ to obtain an exact sequence
$$\mathbb{Z} \otimes \mathbb{Z}/m \to \mathbb{Z} \otimes \mathbb{Z}/m \to \mathbb{Z}/n \otimes \mathbb{Z}/m \to 0$$

C. The tensor product

where the first map is multiplication by n. This translates by the isomorphism in the example above to

$$\mathbb{Z}/m \to \mathbb{Z}/m \to \mathbb{Z}/n \otimes \mathbb{Z}/m \to 0$$

where again the first map is multiplication by n (if n and m are not coprime, the left map is not injective, finishing the proof of Proposition C.2). Thus $\mathbb{Z}/n \otimes \mathbb{Z}/m \cong \mathbb{Z}/\gcd(m,n)$, and we have shown:

Proposition C.3.
$$\mathbb{Z}/n \otimes \mathbb{Z}/m \cong \mathbb{Z}/\gcd(n,m).$$

If A is a finitely generated abelian group it is isomorphic to $F \oplus T$, where $F \cong \mathbb{Z}^k$ is a free abelian group, and T is the torsion subgroup. The number k is called the **rank** of A. A finitely generated torsion group is isomorphic to a finite sum of cyclic groups $\mathbb{Z}/n_i$ for some $n_i > 0$. Thus Propositions C.1 and C.3 allow one to compute the tensor products of arbitrary finitely generated abelian groups.

Now we study the tensor product of an abelian group with the rationals $\mathbb{Q}$. Let A be an abelian group and K be a field. We first introduce the structure of a K-vector space on $A \otimes K$ (where we consider K as an abelian group to construct the tensor product) by: $\alpha \cdot (a \otimes \beta) := a \otimes \alpha \cdot \beta$ for a in A and α and β in K. Decompose $A = F \oplus T$ as above. The tensor product $T \otimes \mathbb{Q}$ is zero, since $a \otimes q = n \cdot a \otimes q/n = 0$, if $n \cdot a = 0$. The tensor product $F \otimes \mathbb{Q}$ is isomorphic to $\mathbb{Q}^k$. Thus $A \otimes \mathbb{Q}$ is — considered as $\mathbb{Q}$-vector space — a vector space of dimension rank A.

Finally we consider an exact sequence of abelian groups

$$\cdots \to A_{k+1} \to A_k \to A_{k-1} \to \cdots$$

and the tensor product with an abelian group P.

Proposition C.4. *Let*

$$\cdots \to A_{k+1} \to A_k \to A_{k-1} \to \cdots$$

be an exact sequence of abelian groups and P either be $\mathbb{Q}$ or a finitely generated free abelian group, then the induced sequence

$$\cdots \to A_{k+1} \otimes P \to A_k \otimes P \to A_{k-1} \otimes P \to \cdots$$

is exact.

Proof: The case of a free finitely generated abelian group P can be reduced to the case $P = \mathbb{Z}$ by Proposition C.1. The conclusion now follows from the isomorphism $A \otimes \mathbb{Z} \cong A$.

If $P = \mathbb{Q}$ we return to Proposition C.2 and note that we are finished if we can show the injectivity of $f \otimes \mathrm{id} : A \otimes \mathbb{Q} \to B \otimes \mathbb{Q}$. Consider an element of $A \otimes \mathbb{Q}$, a finite sum $\sum_i a_i \otimes q_i$, and suppose $\sum_i f(a_i) \otimes q_i = 0$. Let m be the product of the denominators of the q_i's and consider $m(\sum_i a_i \otimes q_i) = \sum_i a_i \otimes m \cdot q_i$. The latter is an element of $A \otimes \mathbb{Z}$ mapping to zero in $B \otimes \mathbb{Q}$. Thus its image in $B \otimes \mathbb{Z}$ is a torsion element (the kernel of $B \cong B \otimes \mathbb{Z} \to B \otimes \mathbb{Q}$ is the torsion subgroup of B (why?)). Since $f \otimes \mathrm{id} : A \otimes \mathbb{Z} \to B \otimes \mathbb{Z}$ is injective, this implies that $\sum_i a_i \otimes m \cdot q_i$ is a torsion element, so it maps to zero in $A \otimes \mathbb{Q}$. Since this is a $\mathbb{Q}$-vector space, $m(\sum_i a_i \otimes q_i) = 0$ implies $\sum_i a_i \otimes q_i = 0$.
q.e.d.

Bibliography

[B-T] R. Bott, L.W. Tu, *Differential forms in algebraic topology.* Graduate Texts in Mathematics, 82, Springer-Verlag, New York-Berlin, 1982.

[B-J] T. Bröcker, K. Jänich, *Introduction to differential topology.* Cambridge University Press, 1982.

[B-R-S] S. Buoncristiano, C.P. Rourke, B.J. Sanderson, *A geometric approach to homology theory.* London Mathematical Society Lecture Note Series, No. 18,. Cambridge University Press, 1976.

[C-F] P.E. Conner, E.E. Floyd, *Differentiable periodic maps. Ergebnisse der Mathematik und ihrer Grenzgebiete.* Neue Folge. 33, Springer-Verlag. 148 p. (1964).

[D] A. Dold, *Geometric cobordism and the fixed point transfer.* Algebraic topology (Proc. Conf., Univ. British Columbia, Vancouver, B.C., 1977), pp. 32–87, Lecture Notes in Math., 673, Springer, Berlin, 1977.

[E-S] S. Eilenberg, N. Steenrod, *Foundations of algebraic topology.* Princeton University Press, Princeton, New Jersey, 1952.

[G] A. Grinberg, *Resolution of Stratifolds and Connection to Mather's Abstract Pre-Stratified Spaces.* PhD-thesis Heidelberg, http://www.hausdorff-research-institute.uni-bonn.de/files/diss_grinberg.pdf, 2003.

[Hi] M. W. Hirsch, *Differential topology.* Springer-Verlag, 1976.

[Hir] F. Hirzebruch, *Topological methods in algebraic geometry.* Third enlarged edition. New appendix and translation from the second German edition by R. L. E. Schwarzenberger, with an additional section by A. Borel. Die Grundlehren der Mathematischen Wissenschaften, Band 131 Springer-Verlag New York, Inc., New York 1966.

[Jä] K. Jänich, *Topology.* Springer-Verlag, 1984.

[K-K] S. Klaus, M. Kreck *A quick proof of the rational Hurewicz theorem and a computation of the rational homotopy groups of spheres.* Math. Proc. Cambridge Philos. Soc. 136 (2004), no. 3, 617–623.

[K-L] M. Kreck, W. Lück, *The Novikov conjecture. Geometry and algebra.* Oberwolfach Seminars, 33, Birkhäuser Verlag, Basel, 2005.

[K-S] M. Kreck, W. Singhof, *Homology and cohomology theories on manifolds.* to appear Münster Journal of Mathematics.

[Mi 1] J. Milnor, *On manifolds homeomorphic to the 7-sphere.* Ann. Math. 64 (1956), 399-405.

[Mi 2] J. Milnor, *Topology from the differentiable viewpoint.* Based on notes by David W. Weaver. The University Press of Virginia, 1965.

[Mi 3] J. Milnor, *Morse theory.* Based on lecture notes by M. Spivak and R. Wells. Annals of Mathematics Studies, No. 51 Princeton University Press, Princeton, N.J. 1963.

[Mi-St] J. Milnor, J. Stasheff, *Characteristic classes.* Annals of Mathematics Studies, No. 76. Princeton University Press, Princeton, N. J.; University of Tokyo Press, Tokyo, 1974.

[Mu] J. Munkres, *Topology: a first course.* Prentice-Hall, Inc., Englewood Cliffs, N.J., 2000.

[Pf] M. Pflaum, *Analytic and geometric study of stratified spaces.* Springer LNM 1768, 2001.

[Po] H. Poincaré, *Analysis situs.* Journal de l'École Polytechnique 1, 1-121, 1895.

[Q] D. Quillen, *Elementary proofs of some results of cobordism theory using Steenrod operations.* Adv. Math. 7 (1970) 29-56.

[Sch] H. Schubert, *Topology.* MacDonald, London, 1968.

[Si] R. Sikorski, *Differential modules.* Colloq. Math. 24, 45-79, 1971.

[S-L] R. Sjamaar, E. Lerman, *Stratified symplectic spaces and reduction.* Ann. Math. (2) 134, 375-422, 1991.

[Th 1] R. Thom, *Quelques propriétés globales des variétés différentiables.* Comment. Math. Helv. 28 (1954) 17-86.

[Th 2] R. Thom, *Ensembles et morphismes stratifiés.* Bull. Amer. Math. Soc. 75 (1969), 240-284.

Index

CW-decomposition, 96
$\mathbb{Z}/2$-Betti number, 74
$\mathbb{Z}/2$-homologically finite, 74
$\mathbb{Z}/2$-homology, 46
$\mathbb{Z}/2$-oriented, 42
$\mathbb{Z}/2$-cohomology groups, 121
⋔, transverse intersection, 125
⌣-product, 135
×-product, 102, 133
c-manifold, 33
c-stratifold, 35
p-stratifold, 24

algebra, 6

Betti numbers, 81
bicollar, 36, 195
Bockstein sequence, 88
bordant, 45
bordism, 45
boundary, 35
boundary operator, 57, 92
bump function, 16

canonical map to tensor product, 208
cells, 96
characteristic class, 151
Chern classes, 160
closed cone, 36
closed manifold, 119
closed unit ball D^n, 67
cohomology ring, 138
cohomology theory, 131
collar, 33, 34
compact c-stratifold, 45
compactly supported homology theory, 93

complex projective space, 73
complex vector bundle, 159
conjugate bundle, 166
connected sum, 13
connective homology theory, 92
contractible space, 50
contravariant functor, 130
covariant functor, 130
cross product, 102, 133
cup product, 135
cutting along a codimension-1 stratifold, 36
cylinder, 35

degree, 83
derivation, 10
differential, 11
differential space, 7

Euler characteristic, 74
Euler class, 129, 151
exact sequence, 55

finite CW-complex, 96
functor, 49, 91, 92
fundamental class, 64, 82
fundamental theorem for finitely generated abelian groups, 81
fundamental theorem of algebra, 84

germ, 9
good atlas, 135
graded commutativity, 136

hedgehog theorem, 85
homologically finite, 81
homology theory, 91

217

homology theory with compact supports, 93
homotopic, 49
homotopy, 49
homotopy axiom, 50
homotopy equivalence, 50
homotopy inverse, 50
Hopf bundle, 168

induced homomorphism in cohomology, 128
induced map, 48
integral cohomology group, 117
integral homology, 80
integral stratifold homology, 80
intersection form, 145
isomorphism, 8, 41

Künneth Theorem for cohomology, 135
Künneth Theorem for homology, 105
Kronecker homomorphism, 140
Kronecker pairing, 140
Kronecker product, 140

lens space, 110
local homology, 69
local retraction, 19
local trivialization, 109

Mayer-Vietoris sequence for homology, 57, 91
Mayer-Vietoris sequence for integral cohomology, 123
Milnor manifolds, 113
morphism, 11

natural equivalence, 93
natural transformation, 88, 92
nice space, 95

one-point compactification, 21
open cone, 20
open unit ball B^n, 67
oriented m-dimensional c-stratifold, 79

parametrized stratifold, 24
partition of unity, 25
path components, 47
path connected space, 47
Poincaré duality, 120
Poincaré duality for $\mathbb{Z}/2$-(co)homology, 121
Pontrjagin classes, 165
Pontrjagin numbers, 168
product formula for Pontrjagin classes, 166
proper map, 117
pure tensors, 207

quaternions, 113

rank of a finitely generated abelian group, 211
rational cohomology, 134
rational homology, 104
reduced homology groups, 61
reduced stratifold homology groups, 80
reduction mod 2, 87
regular c-stratifold, 43
regular stratifold, 43
regular value, 27
relative homology, 70

signature, 146
Signature Theorem, 177
skeleton, 16
smooth fibre bundle, 109
smooth manifold, 9
smooth maps, 25
Stiefel-Whitney classes, 162
stratification, 17
stratifold, 16
stratifold homology, 80
stratifold homology group, 46
stratum, 16
subordinate partition of unity, 25
support of a function, 17

tangent space, 10
tautological bundle, 153
tensor product, 208, 210
top Stiefel-Whitney class, 157
top stratum, 17
topological sum, 23
total Chern class, 161
total Pontrjagin class, 166
transverse, 127
transverse intersection, 125

vector field, 85

Whitney formula, 161, 163

Titles in This Series

115 **Julio González-Díaz, Ignacio García-Jurado, and M. Gloria Fiestras-Janeiro,** An introductory course on mathematical game theory, 2010

114 **Joseph J. Rotman,** Advanced modern algebra: Second edition, 2010

113 **Thomas M. Liggett,** Continuous time Markov processes: An introduction, 2010

112 **Fredi Tröltzsch,** Optimal control of partial differential equations: Theory, methods and applications, 2010

111 **Simon Brendle,** Ricci flow and the sphere theorem, 2010

110 **Matthias Kreck,** Differential algebraic topology: From stratifolds to exotic spheres, 2010

109 **John C. Neu,** Training manual on transport and fluids, 2010

108 **Enrique Outerelo and Jesús M. Ruiz,** Mapping degree theory, 2009

107 **Jeffrey M. Lee,** Manifolds and differential geometry, 2009

106 **Robert J. Daverman and Gerard A. Venema,** Embeddings in manifolds, 2009

105 **Giovanni Leoni,** A first course in Sobolev spaces, 2009

104 **Paolo Aluffi,** Algebra: Chapter 0, 2009

103 **Branko Grünbaum,** Configurations of points and lines, 2009

102 **Mark A. Pinsky,** Introduction to Fourier analysis and wavelets, 2009

101 **Ward Cheney and Will Light,** A course in approximation theory, 2009

100 **I. Martin Isaacs,** Algebra: A graduate course, 2009

99 **Gerald Teschl,** Mathematical methods in quantum mechanics: With applications to Schrödinger operators, 2009

98 **Alexander I. Bobenko and Yuri B. Suris,** Discrete differential geometry: Integrable structure, 2008

97 **David C. Ullrich,** Complex made simple, 2008

96 **N. V. Krylov,** Lectures on elliptic and parabolic equations in Sobolev spaces, 2008

95 **Leon A. Takhtajan,** Quantum mechanics for mathematicians, 2008

94 **James E. Humphreys,** Representations of semisimple Lie algebras in the BGG category $\mathcal{O}$, 2008

93 **Peter W. Michor,** Topics in differential geometry, 2008

92 **I. Martin Isaacs,** Finite group theory, 2008

91 **Louis Halle Rowen,** Graduate algebra: Noncommutative view, 2008

90 **Larry J. Gerstein,** Basic quadratic forms, 2008

89 **Anthony Bonato,** A course on the web graph, 2008

88 **Nathanial P. Brown and Narutaka Ozawa,** C*-algebras and finite-dimensional approximations, 2008

87 **Srikanth B. Iyengar, Graham J. Leuschke, Anton Leykin, Claudia Miller, Ezra Miller, Anurag K. Singh, and Uli Walther,** Twenty-four hours of local cohomology, 2007

86 **Yulij Ilyashenko and Sergei Yakovenko,** Lectures on analytic differential equations, 2007

85 **John M. Alongi and Gail S. Nelson,** Recurrence and topology, 2007

84 **Charalambos D. Aliprantis and Rabee Tourky,** Cones and duality, 2007

83 **Wolfgang Ebeling,** Functions of several complex variables and their singularities (translated by Philip G. Spain), 2007

82 **Serge Alinhac and Patrick Gérard,** Pseudo-differential operators and the Nash–Moser theorem (translated by Stephen S. Wilson), 2007

81 **V. V. Prasolov,** Elements of homology theory, 2007

80 **Davar Khoshnevisan,** Probability, 2007

79 **William Stein,** Modular forms, a computational approach (with an appendix by Paul E. Gunnells), 2007

78 **Harry Dym,** Linear algebra in action, 2007

77 **Bennett Chow, Peng Lu, and Lei Ni,** Hamilton's Ricci flow, 2006

TITLES IN THIS SERIES

76 **Michael E. Taylor,** Measure theory and integration, 2006
75 **Peter D. Miller,** Applied asymptotic analysis, 2006
74 **V. V. Prasolov,** Elements of combinatorial and differential topology, 2006
73 **Louis Halle Rowen,** Graduate algebra: Commutative view, 2006
72 **R. J. Williams,** Introduction the the mathematics of finance, 2006
71 **S. P. Novikov and I. A. Taimanov,** Modern geometric structures and fields, 2006
70 **Seán Dineen,** Probability theory in finance, 2005
69 **Sebastián Montiel and Antonio Ros,** Curves and surfaces, 2005
68 **Luis Caffarelli and Sandro Salsa,** A geometric approach to free boundary problems, 2005
67 **T.Y. Lam,** Introduction to quadratic forms over fields, 2004
66 **Yuli Eidelman, Vitali Milman, and Antonis Tsolomitis,** Functional analysis, An introduction, 2004
65 **S. Ramanan,** Global calculus, 2004
64 **A. A. Kirillov,** Lectures on the orbit method, 2004
63 **Steven Dale Cutkosky,** Resolution of singularities, 2004
62 **T. W. Körner,** A companion to analysis: A second first and first second course in analysis, 2004
61 **Thomas A. Ivey and J. M. Landsberg,** Cartan for beginners: Differential geometry via moving frames and exterior differential systems, 2003
60 **Alberto Candel and Lawrence Conlon,** Foliations II, 2003
59 **Steven H. Weintraub,** Representation theory of finite groups: algebra and arithmetic, 2003
58 **Cédric Villani,** Topics in optimal transportation, 2003
57 **Robert Plato,** Concise numerical mathematics, 2003
56 **E. B. Vinberg,** A course in algebra, 2003
55 **C. Herbert Clemens,** A scrapbook of complex curve theory, second edition, 2003
54 **Alexander Barvinok,** A course in convexity, 2002
53 **Henryk Iwaniec,** Spectral methods of automorphic forms, 2002
52 **Ilka Agricola and Thomas Friedrich,** Global analysis: Differential forms in analysis, geometry and physics, 2002
51 **Y. A. Abramovich and C. D. Aliprantis,** Problems in operator theory, 2002
50 **Y. A. Abramovich and C. D. Aliprantis,** An invitation to operator theory, 2002
49 **John R. Harper,** Secondary cohomology operations, 2002
48 **Y. Eliashberg and N. Mishachev,** Introduction to the h-principle, 2002
47 **A. Yu. Kitaev, A. H. Shen, and M. N. Vyalyi,** Classical and quantum computation, 2002
46 **Joseph L. Taylor,** Several complex variables with connections to algebraic geometry and Lie groups, 2002
45 **Inder K. Rana,** An introduction to measure and integration, second edition, 2002
44 **Jim Agler and John E. McCarthy,** Pick interpolation and Hilbert function spaces, 2002
43 **N. V. Krylov,** Introduction to the theory of random processes, 2002
42 **Jin Hong and Seok-Jin Kang,** Introduction to quantum groups and crystal bases, 2002
41 **Georgi V. Smirnov,** Introduction to the theory of differential inclusions, 2002
40 **Robert E. Greene and Steven G. Krantz,** Function theory of one complex variable, third edition, 2006
39 **Larry C. Grove,** Classical groups and geometric algebra, 2002

For a complete list of titles in this series, visit the
AMS Bookstore at www.ams.org/bookstore/.

DATE DUE

QA 612 .K74 2010

Kreck, Matthias, 1947-

Differential algebraic
 topology